T0143398

Electronic System-Level HW/SW Co-Design of Heterogeneous Multi-Processor Embedded Systems

RIVER PUBLISHERS SERIES IN CIRCUITS AND SYSTEMS

Volume 2

Series Editors

MASSIMO ALIOTO
National University of Singapore
Singapore

KOFI MAKINWA
Delft University of Technology
The Netherlands

DENNIS SYLVESTER
University of Michigan
USA

The "River Publishers Series in Circuits & Systems" is a series of comprehensive academic and professional books which focus on theory and applications of Circuit and Systems. This includes analog and digital integrated circuits, memory technologies, system-on-chip and processor design. The series also includes books on electronic design automation and design methodology, as well as computer aided design tools.

Books published in the series include research monographs, edited volumes, handbooks and textbooks. The books provide professionals, researchers, educators, and advanced students in the field with an invaluable insight into the latest research and developments.

Topics covered in the series include, but are by no means restricted to the following:

- Analog Integrated Circuits
- Digital Integrated Circuits
- Data Converters
- Processor Architecures
- System-on-Chip
- Memory Design
- Electronic Design Automation

For a list of other books in this series, visit www.riverpublishers.com

Electronic System-Level HW/SW Co-Design of Heterogeneous Multi-Processor Embedded Systems

Luigi Pomante

Center of Excellence DEWS
Università degli Studi dell'Aquila
L'Aquila, Italy

River Publishers

Published, sold and distributed by:
River Publishers
Alsbjergvej 10
9260 Gistrup
Denmark

River Publishers
Lange Geer 44
2611 PW Delft
The Netherlands

Tel.: +45369953197
www.riverpublishers.com

ISBN: 978-87-93379-38-1 (Hardback)
   978-87-93379-37-4 (Ebook)

©2016 River Publishers

All rights reserved. No part of this publication may be reproduced, stored in a retrieval system, or transmitted in any form or by any means, mechanical, photocopying, recording or otherwise, without prior written permission of the publishers.

To Beatrice, with Love

Contents

Preface   xi

Acknowledgments   xiii

List of Figures   xv

List of Tables   xix

List of Abbreviations   xxi

Part 1: System-Level Co-Design of Heterogeneous
Multi-Processor Embedded Systems

1 Introduction   3

2 Background   7
  2.1 Heterogeneous Multi-Processor Embedded Systems   7
      2.1.1 Existing Projects   11
      2.1.2 Design Issues   12
  2.2 Concurrent HW/SW Design   13
      2.2.1 State-of-the-Art   18
  2.3 Conclusion   25

3 The Proposed Approach   27
  3.1 The Reference Environment: TOSCA   27
      3.1.1 The Specification Language   28
      3.1.2 Intermediate Representations   29
      3.1.3 The Target Architecture   30
      3.1.4 Overview of the Design Flow   32
  3.2 The Proposed Environment: TOHSCA   39
      3.2.1 Target Architecture   44
  3.3 Conclusion   46

4   System-Level Co-Specification   **47**
    4.1   System-Level Specification Languages   48
    4.2   Reference Language   51
          4.2.1   OCCAM   51
    4.3   Internal Models   54
          4.3.1   Statement-Level Internal Model   55
          4.3.2   Procedure-Level Internal Model   57
    4.4   Conclusion   60

5   Metrics for Co-Analysis   **63**
    5.1   Characterization 65
          5.1.1   GPP Architectural Features   66
          5.1.2   DSP Architectural Features   66
          5.1.3   ASIC-like Devices Architectural Features   69
    5.2   The Proposed Approach   70
          5.2.1   Model and Methodology   72
          5.2.2   The Tool   81
          5.2.3   Validation   82
    5.3   Conclusion   83

6   System-Level Co-Estimations   **85**
    6.1   Characterization 87
          6.1.1   Performance Estimation   87
    6.2   The Proposed Approach   88
          6.2.1   Model and Methodology   88
          6.2.2   Application of the model to OCCAM2   97
          6.2.3   The Tool 117
          6.2.4   Validation   119
    6.3   Conclusion 122

7   System-Level Partitioning   **123**
    7.1   Characterization   124
    7.2   The Proposed Approach 126
          7.2.1   Model and Methodology   127
          7.2.2   The Tool 134
          7.2.3   Validation   135
    7.3   Conclusion 138

8   System-Level Co-Simulation   **139**
    8.1   Characterization   141
    8.2   The Proposed Approach 142

	8.2.1	Model and Methodology	143
	8.2.2	The Tool	152
	8.2.3	Validation	155
8.3		Conclusion	158

9 Case Studies   **159**
9.1		Case Study 1	159
	9.1.1	Co-specification	159
	9.1.2	Co-analysis	160
	9.1.3	Co-estimation	161
	9.1.4	Functional Co-simulation	162
	9.1.5	Load Estimation	163
	9.1.6	System Design Exploration	163
	9.1.7	Toward the Low-level Co-design Flow	165
9.2		Case Study 2	165
	9.2.1	Co-specification	166
	9.2.2	Co-analysis	166
	9.2.3	Co-estimation	167
	9.2.4	Functional Co-simulation	167
	9.2.5	Load Estimation	168
	9.2.6	System Design Exploration	168
	9.2.7	Toward a Low-level Co-design Flow	170
9.3		Conclusion	170

Conclusions (Part 1)   **171**

Part 2

10 System-Level Design Space Exploration   **175**
10.1	Introduction	175
10.2	Reference Co-Design Flow	177
10.3	Specification	181
10.4	Target HW Architecture	185
10.5	Design Space Exploration	187
	10.5.1 First Phase	187
	10.5.2 Second Phase	192
	10.5.3 Illustrative Example	195
10.6	Conclusion	198

11 SystemC-Based ESL Design Space Exploration   **199**
  11.1 Introduction 199
  11.2 Reference ESL HW/SW Co-Design Flow 201
      11.2.1 System Behavior Model 201
      11.2.2 Technologies Library 203
      11.2.3 Functional Simulation 203
      11.2.4 Co-Analysis and Co-Estimation 204
      11.2.5 Design Space Exploration 204
      11.2.6 Algorithm-Level Flow 205
      11.2.7 Reference Template HW Architecture 205
  11.3 SystemC-Based ESL HW/SW Co-Design Environment . . . 206
      11.3.1 System Behavior Model 206
      11.3.2 Functional Simulation 207
      11.3.3 Co-Analysis and Co-Estimation 207
      11.3.4 Design Space Exploration 208
  11.4 SystemC-Based ESL Design Space Exploration 208
      11.4.1 HW/SW Partitioning, Mapping, and Architecture
             Definition (1st Phase) 209
             11.4.1.1 Inputs modeling 210
             11.4.1.2 Technologies library modeling 211
             11.4.1.3 PAM specification modeling 211
             11.4.1.4 Optimization engine, individuals,
                     and allocation modeling 213
      11.4.2 Timing Co-Simulation 213
  11.5 FIR-FIR-GCD Case Study 216
  11.6 Conclusion 222

References   **223**

Index   **239**

About the Author   **245**

Preface

This book presents the research activities that have started with my PhD course and their evolution during the following (more than ten) years. For this, the book is divided into two parts.

The first part (i.e. Part 1), called *"System-Level Co-design of Heterogeneous Multi-Processor Embedded Systems"*, presents a slight revisitation of my PhD Thesis. Fortunately, other than representing a starting point to fully understand the second part, its content is still very actual.

Then, the second part (i.e. Part 2) presents the main outcomes of the research acitivites that I have performed after my PhD on the same topic (it is worth noting that I had to work also on other, more or less related, ones). In this part, the main focus is on *Design Space Exploration* and it is composed of two "near self-contained" chapters (i.e. they propose again some key concepts from Part 1 and partially overlap). The first one, called *"System-Level Design Space Exploration for Dedicated Heterogeneous Multi-Processor Systems"*, is about main theoretical advancements with respect to Part 1; the second one, called *"SystemC-based ESL Design Space Exploration for Dedicated Heterogeneous Multi-Processor Systems"*, presents a prototypal framework that provides practical exploitation of the proposed concepts.

The title of the whole book (*"Electronic System-Level HW/SW Co-Design of Heterogeneous Multi-Processor Embedded Systems"*) tries to represent the two parts in a unified way.

Acknowledgments

I would like to thank all the people who have actively contributed to the research presented in this book and who have supported my work during my PhD course and the following researcher career. In particular, I am especially thankful to my mentor Prof. Donatella Sciuto and my colleagues Prof. Fabio Salice, Prof. William Fornaciari, Dr. Carlo Brandolese, Prof. Cristiana Bolchini, and Ing. Alberto Allara for the valuable experience and the unconstrained help that they have provided to me when I was at Politecnico di Milano and CEFRIEL. I am thankful also to my colleagues Prof. Fortunato Santucci, Prof. Fabio Graziosi, Prof. Vittorio Cortellessa, Prof. Marco Faccio, and Prof. Maria Domenica Di Benedetto for the opportunity, the help, and the collaboration that they have provided to me when I moved from Politecnico di Milano to Center of Excellence DEWS (Università degli Studi dell'Aquila). Moreover, I would like to thank all the students I had the opportunity to work with on the topics of this book and that have been a valuable support in many occasions. I would especially like to mention Azzurra Persico, Aldo Vico, Luca Del Vecchio, Laura Imbriglio, Paolo Serri, and Jacopo Volpe. Finally, I would like to thank my wife Beatrice, my parents, my brother, and all the friends, more or less intimate, who have been close to me during my career and life: their names are clear and bright in my heart but, fortunately, are too many to be listed here.

Thanks,
Luigi

List of Figures

Figure 2.1	Modern co-design framework.	14
Figure 2.2	Traditional top-down flow vs. co-design flow. . . .	17
Figure 2.3	Virtual prototype based framework.	19
Figure 3.1	Outline of the TOSCA co-design flow.	28
Figure 3.2	The VIS language syntax.	30
Figure 3.3	The TOSCA target architecture.	31
Figure 3.4	The TOSCA co-processor internal structure.	31
Figure 3.5	The TOSCA compilation flow.	34
Figure 3.6	Back-annotation and cross-references.	39
Figure 3.7	The proposed high-level flow.	40
Figure 3.8	Target architecture.	45
Figure 4.1	Structure of the OCCAM internal model.	56
Figure 4.2	Structure of PROC internal model.	56
Figure 4.3	Structure of PAR, SEQ, IF, and ALT internal models.	56
Figure 4.4	Structure of a syntax tree.	57
Figure 4.5	Internal model graphical representation.	60
Figure 5.1	The proposed high-level flow.	64
Figure 5.2	SHARC Internal Architecture.	67
Figure 5.3	MAC example.	68
Figure 5.4	A circular buffer of length k.	68
Figure 5.5	Bit reversing in the FFT.	69
Figure 5.6	Examples of circular buffering.	75
Figure 5.7	Normalization function.	81
Figure 6.1	The proposed high-level flow.	86
Figure 6.2	Sample code decomposition: (a) Source, (b) hierarchy, and (c) execution paths.	89
Figure 6.3	Sample mapping to target assembly language. . . .	94
Figure 6.4	The considered model (a) and an example of algebraic expression graph (b).	100
Figure 6.5	Syntax of the IF statement.	102
Figure 6.6	Translation template for the IF statement.	103

Figure 6.7   Syntax of the WHILE statement. 104

Figure 6.8   Translation template for the WHILE statement. . . 104

Figure 6.9   Data structure for channels implementation. 105

Figure 6.10   Syntax of the ALT statement. 107

Figure 6.11   Three-children SEQ template. 111

Figure 6.12   Three-children PAR template. 112

Figure 6.13   Three-children IF template. 113

Figure 6.14   WHILE template. 114

Figure 6.15   Communication template. 114

Figure 6.16   Channel state diagram. 115

Figure 6.17   A complete rendezvous. 115

Figure 6.18   Two-children ALT template. 116

Figure 6.19   CPI and operator count data structure. 118

Figure 6.20   Tree model of the assignment $Y:=4*(A+B+1)$. . . 119

Figure 6.21   Operator and variable counts data structure. 119

Figure 7.1   The proposed high-level flow. 124

Figure 7.2   Partitioning methodology steps. 131

Figure 7.3   Procedural interaction graph. 132

Figure 7.4   Individual structure. 133

Figure 7.5   Crossover. 133

Figure 7.6   Procedure Interaction Graph. 135

Figure 8.1   The proposed high-level flow. 140

Figure 8.2   Notations. 145

Figure 8.3   An example of the time-stretching procedure
for three processes. 147

Figure 8.4   Multi-processors time-stretching algorithm. . . . 148

Figure 8.5   Starting communications management. 151

Figure 8.6   Ending communications management. 152

Figure 8.7   The functional composition of the target
example. 155

Figure 8.8   The multimedia application: procedure
call graph. 156

Figure 8.9   Performance vs. Architecture. 157

Figure 9.1   The proposed system-level co-design flow. 160

Figure 9.2   Procedure interaction graph. 161

Figure 9.3   The VCG file. 164

Figure 9.4   Procedure interaction graph. 166

Figure 10.1   The reference system-level co-design flow. 178

Figure 10.2   Typical algorithm-level design flow. 182

Figure 10.3	CSP and a related PING.	184
Figure 10.4	The Basic Block and its characterization parameters.	186
Figure 10.5	Architecture graph.	187
Figure 10.6	The two-phase DSE approach.	188
Figure 10.7	Procedural interaction graph.	189
Figure 10.8	An individual and its corresponding architecture.	191
Figure 10.9	Crossover.	191
Figure 10.10	BBs interaction graph.	192
Figure 10.11	An example of individual and corresponding architecture.	194
Figure 10.12	Cross-over.	195
Figure 10.13	CU characterization matrix.	195
Figure 10.14	Best individual and corresponding architecture. . .	197
Figure 10.15	Best individual 2 and the corresponding architecture.	197
Figure 11.1	The reference co-design flow.	202
Figure 11.2	CSP and a related PING.	203
Figure 11.3	SC_CSP_CHANNEL interface and *read()* implementation.	207
Figure 11.4	An example of System and Test-Bench modeling.	208
Figure 11.5	UML model related to the annotated PING XML schema.	210
Figure 11.6	Technologies library modeling.	211
Figure 11.7	PAM specification modeling.	212
Figure 11.8	Individuals and allocation modeling.	214
Figure 11.9	Optimization engine modeling.	215
Figure 11.10	SystemC HW/SW Timing Co-Simulator architecture.	215
Figure 11.11	CSP representing the SBM.	217
Figure 11.12	Sketches of SystemC descriptions of the main SC_MODULE and a CSP process.	217
Figure 11.13	Sketch of SystemC *main()*.	218
Figure 11.14	Outputs from functional validation.	218
Figure 11.15	Affinity with respect to GPP, DSP, and SPP.	219
Figure 11.16	Worst-case time-to-completion.	220
Figure 11.17	DSE step results.	221

List of Tables

Table 3.1   OCCAM subset supported by TOSCA   29
Table 4.1   OCCAM subset supported by TOSCA   53
Table 4.2   OCCAM extension for constraints support   53
Table 5.1   Different affinity average values   83
Table 6.1   The defined set of instruction classes   97
Table 6.2   Models for software assignments   99
Table 6.3   Models for software arithmetic expressions
             (minimum values)   100
Table 6.4   Models for software arithmetic expressions
             (maximum values)   100
Table 6.5   Models for software logic expressions
             (best case)   101
Table 6.6   Models for software logic expressions
             (worst case)   102
Table 6.7   Timing model for the IF control statement   103
Table 6.8   Timing model for the WHILE   104
Table 6.9   Timing model for the input process   106
Table 6.10   Timing model for the output process   107
Table 6.11   Timing model for the guarded-input process   108
Table 6.12   Timing model for the ALT arbiter   108
Table 6.13   Models for hardware assignments   109
Table 6.14   Models for hardware arithmetic expressions   110
Table 6.15   Timing performance of the main hardware operators
             implementations   110
Table 6.16   Timing performance of the hardware logical
             and relational operators implementations   111
Table 6.17   Sample CPI file for the Intel i80486DX2
             processor   118
Table 6.18   Static timing estimates for the GCD algorithm . .   120
Table 6.19   Dynamic timing estimates, measures and errors
             for the GCD algorithm   121

Table 6.20   Static timing estimates for the BSORT
algorithm 121
Table 6.21   Dynamic timing estimates, measures and errors
for the BSORT algorithm 122
Table 7.1   Affinity values 136
Table 7.2   Cost function weights 137
Table 7.3   Timing constraint: 90% T_{REF} 137
Table 7.4   Timing constraint: 50% T_{REF} 137
Table 8.1   Channel Communication times. 149
Table 8.2   Procedure call times 150
Table 8.3   Experimental results 157
Table 9.1   Affinity values 162
Table 9.2   Design space exploration 164
Table 9.3   Affinity values 167
Table 9.4   Design space exploration 169

List of Abbreviations

aLICE	a Library for Integrated Co-design Environment
AES	Advanced Encryption Standard
ARM	Advanced RISC Machines
ASIC	Application Specific Integrated Circuits
ASIP	Application Specific Instruction Processor
ASP	Application Specific Processor
BB	Building Block
BING	Building Blocks Interaction Graph
BSORT	Bubble SORT
CAD	Computer Aided Design
CC	Concurrent Communications
CDFG	Control Data Flow Graph
CF	Cost Function
CHAMP	Common Heterogeneous Architecture for Multi-Processing
CIS	Compatible Instruction Set
CISC	Complex Instruction Set Computer
COTS	Commercial (Common) Off The Shelf
CPI	Clock cycles Per Instruction
CRC	Cyclical Redundancy Check
CSP	Communicating Sequential Processes
CU	Communication Unit
CUCM	Communication Units Characterization Matrix
DAG	Data Address Generator
DAG	Dyrected Acyclic Graph
DEVKIT	DEVelopment KIT
DRAM	Dynamic Random Access Memory
DS	Dedicated System
DSE	Design Space Exploration
DSP	Digital Signal Processor
ECU	External Communication Unit
EDA	Electronic Design Automation

EIL	External Interconnection Link
EMUP	Embedded Multi-processor Partitioning
ESL	Electronic System Level
FIFO	First In First Out
FIR	Finite Impulse Response
FFT	Fast Fourier Transform
FPD	Field Programmable Devices
FPGA	Field Programmable Gate Array
GA	Genetic Algorithm
GALIB	Genetic Algorithms LIBrary
GCD	Greatest Common Divisor
GPIO	General Purpose Input Output
GPP	General Purpose Processor
GPR	General Purpose Register
GPU	Graphical Processing Unit
HDL	Hardware Description Language
HMPS	Heterogeneous Multi Processor System
HW	Hardware
IC	Integrated Circuit
I^2C	Inter Integrated Circuit
ICE	Intrinsic Computational Efficiency
IIL	Internal Interconnection Link
ILP	Instruction Level Parallelism
IP	Industrial Property
ISA	Instruction Set Architecture
JPEG	Joint Photographic Experts Group
KBC	KByte for Code
KBD	KByte for Data
KLOC	Kilo Lines Of Code
LM	Local Memory
LSI	Large Scale of Integration
LUT	Look Up Table
MAC	Multiply Accumulate
MD5	Message Digest 5
MIMD	Multiple Instructions Multiple Data
MIPS	Microprocessor without Interlocked Pipeline Stages
MOC	Model Of Computation
MPEG	Moving Picture Experts Group
MPU	Micro Processing Unit

NOO	Not Object Oriented
NOP	No OPeration
OBDD	Ordered Binary Decision Diagram
OO	Object Oriented
OSTE	Occam Simulator for the TOSCA Environment
PAM	Partitioning, Architecture definition and Mapping
PING	Procedural INteraction Graph
PU	Processing Unit
RAM	Random Access Memory
RISC	Reduced (Restricted) Instruction Set Computer
ROM	Read Only Memory
RT	Register Transfer
RTL	Register Transfer Level
SBM	System Behaviour Model
SDL	System Description Language
SHARC	Super Harvard ARChitecture
SLDL	System Level Description Language
SLET	System-Level Estimation Tool
SOC	System On Chip
SPARC	Scalable Processor ARChitecture
SPI	Serial Peripheral Interface
SPP	Single (Specific) Purpose Processor
SW	Software
TL	Technology Library
TO(H)SCA	TOols for Heterogeneous multi-processor embedded Systems Co-design Automation
TOSCA	TOols for System Co-design Automation
TTC	Time To Completion
TV	TeleVision
UART	Universal Asynchronous Receiver Transmitter
UML	Unified Modeling Language
VHDL	VHSIC (Very High Speed Integrated Circuits) Hardware Description Language
VIS	Virtual Instruction Set
VCG	Visualization and Computer Graphics
WCET	Worst Case Execution Time
WCTTC	Worst Case Time To Completion
XML	eXtensible Markup Language

Part 1

November 1998–December 2001

System-Level Co-Design
of Heterogeneous
Multi-Processor Embedded Systems

1

Introduction

Modern electronic systems consist of a fairly heterogeneous set of components. Today, a single system can be constituted by a hardware platform, frequently composed of a mixture of analog and digital components, and by several software application layers. The hardware can include several microprocessors (general purpose, *DSP, or GPU*), dedicated *ICs* (ASICs and/or FPGAs), memories, a set of local connections between the system components, and some interfaces between the system and the environment (sensors, actuators, etc.).

Therefore, on one hand, multi-processor embedded systems seem to be capable to meet the demand of processing power and flexibility of complex applications. On the other hand, such systems are very complex to design and optimize, so that the design methodology plays a major role in determining the success of the products. For these reasons, to cope with the increasing system complexity, the approaches typically used in current projects are oriented toward a co-design methodology working at the higher levels of abstraction. Unfortunately, such methodologies are typically *customized* for the specific application, suffer from the lack of generality, and still need a considerable effort when a real-size project is envisioned.

Therefore, there is the need for a general methodology capable of supporting the designer during the high-level steps of a co-design flow, enabling an effective design space exploration before tackling the low-level steps and thus committing to the final technology. This should prevent costly redesign loops.

The work described in this part of the book (i.e., Part 1) aims at providing models, methodologies, and tools to support each step of the co-design flow of embedded systems implemented by exploiting heterogeneous multi-processor architectures mapped on distributed systems, as well as fully integrated onto a single chip. The result is a significant extension of an existing single-processor HW/SW co-design environment, in order to support,

3

at system-level, multi-processor embedded systems HW/SW co-design. In particular, Part 1 focuses on the following issues:

- analysis of system specification languages, and development of an intermediate representation for the reference one;
- definition of innovative metrics for the analysis of the system specification in order to statically detect the most appropriate processing element for each system functionality;
- analysis and extension of an existing system-level HW/SW performance estimation methodology;
- development of an innovative system-level HW/SW partitioning methodology, supporting heterogeneous multi-processor architecture selection; and
- analysis and extension of an existing system-level HW/SW co-simulation methodology to support heterogeneous multi-processor architectures, considering a high-level model for the communication media.

Part 1 is organized as follows:

Chapter 2 presents the main concepts related to heterogeneous multi-processor embedded systems, describing their general architecture and their application fields, and identifying the criticalities of their design. With the purpose of describing a possible design strategy for such systems, the HW/SW co-design approach and its main issues are introduced, showing the advantages with respect to classical approaches and discussing the implications of its extension to cover heterogeneous multi-processor systems.

Chapter 3 presents the starting point of this project, the *TOSCA* environment (an existing single-processor HW/SW co-design environment) and the extensions proposed in this book (*TOHSCA*) to support HW/SW co-specification, HW/SW co-analysis, HW/SW co-simulation, and system design exploration of heterogeneous multi-processor embedded systems.

Chapter 4 presents a review of the *state-of-the-art* formalisms used for system-level specification. Then, the *TOHSCA* reference language and the internal models used to represent the specification are introduced. In particular, the chapter shows how the procedural-level internal model, defined and adopted in this book, is suitable to represent the main features of several specification languages enabling the proposed system design exploration methodology to be adopted with different types of specification.

Chapter 5 addresses the definition of a set of metrics, providing quantitative information useful to take system-level decisions such as architectural selection and HW/SW partitioning. The underlying idea is that the performance metrics of a final design can be related to the properties of the specification itself. Therefore, the core of this step involves the identification and the evaluation of functional and structural properties of specification, which could affect design performance on different architectural platforms. The proposed metrics expresses the *affinity* of functionality toward each possible processing element (*GPP*, *DSP*, and *ASIC/FPGA*), data that are then considered during the system design exploration step.

Chapter 6 addresses the problem of estimating software and hardware performance at a high-level of abstraction, necessary to enable design space exploration, while maintaining an acceptable level of accuracy. The proposed methodology is general enough to be applicable to several formalisms and co-design environments. It is based on uniform modeling of the system components, where the performance of both hardware and software is expressed in terms of *CPI*, and specific techniques to estimate such values starting from high-level specifications are discussed.

Chapter 7 describes the *"partitioning and architecture selection"* and *"timing co-simulation"* compose the system design exploration step of the proposed HW/SW co-design flow. More in detail, after introducing the main partitioning issues, the metrics and the cost function adopted in the proposed approach are accurately defined showing the interaction with the other tools of the environment. In the following, the methodology, based on an initial *clustering* and on a heuristic optimization step, is analyzed in detail.

Chapter 8 introduces a modeling approach and the related simulation strategy, to represent the behavior of multi-processor HW/SW architectures starting from system-level specifications. The co-simulation kernel is encapsulated within the TOHSCA co-design toolset and interfaced with the software suite computing the evaluation metrics driving the user during the partitioning task. The proposed approach is particularly valuable since it allows the designer to maintain the analysis at a very abstract level, while gathering significant information on the hypothetical architecture of the final system implementation.

Chapter 9 shows the effectiveness and the efficiency of the proposed flow by describing two case studies. In particular, each step of the flow is considered in detail, describing its role in the environment and the data exchanged with the other tools. Moreover, the case studies focus on the tools used and their interaction in order to emphasize the operative issues.

Finally, the conclusions summarize the main contributions of this part of the book and analyze the future developments of the co-design methodology. Some of such developments have been addressed later in the work described in Part 2.

2

Background

This chapter briefly presents the main features of a heterogeneous multi-processor embedded architecture, to highlight the potential benefits over a more conventional hw/sw solution. The basic characteristics have been identified and organized to provide the reader with an exhaustive taxonomy, useful to discover the proper ranges of applicability and to put in evidence the main tradeoffs the designer have to face with. Moreover, an analysis of some representative examples/projects, coming from both commercial and academic fields, to show how such *abstract* features can be really found in real-world systems has been included.

The final part of the chapter is devoted to present the characteristics of the *co-design* discipline and the peculiarity of its application whenever multi-processor embedded systems are the target implementation platform. A review of the related co-design environments and proposals, which appeared in the meaningful literature, is also included.

2.1 Heterogeneous Multi-Processor Embedded Systems

Heterogeneous multi-processor architectures are common for those systems in which various kinds of communication links interconnect programmable and application-specific processing elements and memories, and multiple tasks are concurrently executed. In fact, *heterogeneous multi-processor embedded systems* may consist of *general-purpose* processors (*GPP*), micro-controllers, *digital-signal* processors (*DSP*), *application-specific instructions* processors (*ASIP*), *application-specific* processors (*ASP*), *application-specific* integrated circuits (*ASIC*), and *field-programmable gate arrays* (*FPGA*), and memory modules, properly interconnected by some kind of network topology (point-to-point, bus, multiple buses, mesh, etc.) to perform application-specific functions. Finally, depending on the application field, the system can be implemented on a single chip (SOC) or on one or more (also distributed)

7

boards (*board-level design*) exploiting the reuse of available components (IP cells and microprocessor cores for SOC, COTS components for systems on board) in order to reduce costs and time-to-market.

There are many possible reasons for the choice of a similar architecture for an embedded system. Often, there are conflicting goals, and trade-offs have to be made to find the best compromise between them. The main aspects typically considered [3, 83], in the trade-off analysis, are briefly introduced as follows:

Performance

The primary goal of the system design process is to find a implementation of the system able to meet all the deadlines. For some applications, there might not be any processors available with sufficient performance, and therefore, it becomes necessary to exploit possible parallelism or to use ASICs to gain enough computational power.

However, it has been shown [1], based on the results obtained by numerous researchers and on theoretical grounds provided by *Amdahl's Law* [2] (which limits the possible overall speedup effects obtainable by accelerating a fraction of a program), that it is not possible to reach an arbitrary speed-up for general programs by moving parts of it to the hardware. There are of course some applications (e.g., some digital signal-processing ones), where a small portion of the program contributes to almost the entire execution time, but such situations are less frequent when the system complexity increases. Therefore, in many cases, an application-specific heterogeneous multi-processor architecture is needed in order not only to exploit the explicit or implicit parallelism present in the application by means of the computational power provided by several cooperating processing elements, but also to tailor the system to the application needs. More in detail, with such an architecture, it is possible:

- to use the most appropriate processing element to perform a specific functionality (e.g., DSPs for signal-processing tasks and GPPs for control tasks);
- to allocate tasks with different timing characterizations separately (e.g., periodic and sporadic tasks, tasks with hard and soft real-time constraints, etc.) in order to select the most appropriate local scheduling policy; and
- to minimize communications by allocating cooperating tasks in the same subsystem.

Predictability

One of the most important tasks of a real-time system is to produce correct logical results as dictated by the functional specification on the system. Thus, fast computational speed is a valuable aid in producing correct results faster but it does not constitute a guarantee for correct system behavior since, in real-time systems, the correctness of the results and the exact time at which the results are produced are of equal importance [77]. Therefore, to guarantee the correctness (functional and timing) of high-performance real-time applications, a multi-processor architecture with predictable behavior is often a necessity.

The predictability property for a computer system indicates the amount of determinism in its temporal behavior while working under timing constraints [78]. Such property is then the essential factor for real-time systems, embedded or not, and it depends heavily on the predictability of the lower functional levels of the computer hierarchy. Hardware with deterministic behavior provides a powerful basis for the design of deterministic low-level software mechanisms, which in turn constitute a framework for construction of predictable software applications. Similarly, knowledge about the degree of predictability in each hardware and software layer facilitates timing analysis of the system real-time properties.

Modern general processors are equipped with mechanisms that have emerged in order to enhance the average performance of the processor itself. Unfortunately, these mechanisms cannot be used freely in a system whose behavior should be predictable. In fact, mechanisms like pipelining (data and control hazards affect performance as well as predictability analysis) and cache memories (cache hit and cache miss situations affect timing operations in an unpredictable way) are often ignored in timing analysis of real-time system architectures. Only if reasonably accurate worst-case analysis is performed on the architecture, certain features can be allowed on the system [81, 82].

The same consideration applies for RISC processors, heavily based on pipelining and caching techniques in order to maintain a high performance [79, 80] and, in general, for each kind of advanced *Instruction Level Parallelism* (*ILP*) processors (*super-pipelined* processors, *super-scalar processors*, etc.).

A different situation arises instead for *Digital Signal Processors* (*DSP*). In fact, they are microprocessors that can handle different operations when compared with conventional processors and, typically, DSPs can manipulate continuous data flowing in real time. They perform a single task with minimal latency and with limited memory and peripheral devices use offering, for particular tasks, good performance with a considerable degree of predictability.

These considerations lead to the fact that heterogeneous architectures, constituted by several *simple* GPPs (for control flows management) and DSPs (for data flows management), are an optimal candidate as target architectures for real-time systems that need high performance and good behavior predictability.

Cost

To construct ASICs or to use high-end microprocessors is usually quite expensive, so (other than for predictability issues) a cost-effective way to build a system is represented by the use of several cheaper processors that work concurrently to provide the needed computational power. Such approach, therefore, represents the first choice for complex systems, considering that it is also possible to use IP cells, microprocessors cores, or COTS component to reduce the design costs, leaving the use of dedicated hardware as an ultimate choice for critical sections.

Flexibility

If a part of the behavior is likely to be modified after the system is in operation, or if several versions of the same system are planned, it is important to allow changes to be made as easily as possible. This is an argument for choosing to implement the parts that are expected to be modified in the software while adopting, however, the most appropriate processing elements in order to avoid performance depreciation. Finally, for critical functionalities, it is also possible to consider a reconfigurable executor (e.g., FPGA) in order to keep both flexibility and performance.

Distribution

In some cases, the use of a heterogeneous multi-processor architecture is dictated by the environment. For instance, if the sensors and actuators of the system are geographically dispersed, a distributed architecture is a constraint, to place the computing resources close to the related parts of the environment.

Weight

Opposite to the distribution issues, many embedded systems need to be in some sense portable, and then the weight (other than the power consumption!) of the implementation becomes important. This is, for instance, the case for mobile telephones, and also aerospace and automotive applications. If weight is an issue, it can be desirable to integrate as much functionalities as possible into a few chips, to reduce the number of cards and external buses, still keeping the needed performance: a heterogeneous multi-processor system integrated in a single chip offers the best opportunity to reach these goals.

Fault tolerance

Most embedded systems are safety-critical so they must operate, at least partially, even under severe disturbances. Therefore, it might be necessary to replicate some specific functionalities by allocating them on different processing elements, or implement the same part with different technologies to reduce the risk of systematic errors that could affect a certain kind of components in particular environmental conditions. Heterogeneous multi-processor architectures are then an optimal choice for this kind of issues.

Power consumption

Heterogeneous multi-processor systems could also help to overcome one of the major current bottlenecks with respect to the performance of embedded systems, that is, the power consumption. By processing as much as possible in parallel, exploiting simple processors instead of advanced architecture (e.g., *superscalar*, *superpipelined*, etc.), the clock frequencies and the voltages, and therefore the power consumption, can be kept at acceptable levels [107].

The list is obviously not exhaustive and many other aspects could be considered for particular applications.

2.1.1 Existing Projects

Heterogeneous multi-processor embedded systems have been exploited for the implementation of different applications both for research and for commercial use. Let us briefly introduce some representative examples.

In [107], a heterogeneous multi-processor system on chip for the management of real-time video streams has been developed based on an *ad hoc* design methodology. The result is a chip that can manage up to 25 internal real-time video streams. The chip combines the flexibility of a programmable solution with the cost effectiveness of a consumer product.

A similar example is described in [108], where a single-chip multi-processor system for video signal-processing applications has been developed integrating four processing nodes with *on-chip DRAM* and application-specific interfaces. The work focuses on the methodologies developed to design a single node of the system (*AxPe* processor core) dealing with methods for an efficient use of the integrated memory. A final example presents the implementation of a real-time *MPEG2* encoder/decoder using two multi-processor chips.

The concept of *Intrinsic Computational Efficiency* (*ICE*, i.e., number of operations per second per watt) has been defined in [83] and used to show the advantages of heterogeneous multi-processor architectures in the field of embedded systems. Moreover, the work shows two *off-the-shelf* examples from different application domains (digital *TV* and intelligent telephone terminal) in which a GPP and a DSP (each one with its own operating system) interconnected by a proper medium, has been adopted as the optimal implementation mix.

Finally, heterogeneous multi-processor embedded systems are available also commercially, where several vendors of customizable and scalable *system-on-board* offer heterogeneous multi-processor architecture systems as building blocks for complex embedded systems. A representative example is the *Ixthos CHAMP* (*Common Heterogeneous Architecture for Multi-Processing*, [109]), while others can be found on several vendor sites (e.g., *Mercury Computer Systems* [110], *Alacron* [111], *Sky Computer* [112], etc.). Moreover, each vendor provides different development environments targeted to the proposed architecture.

2.1.2 Design Issues

Multi-processor embedded systems are a promising solution for a broad range of modern and complex applications. However, their design complexity and management is relevant, and no assessed design methodology is available today. In fact, for all the examples previously cited, *ad hoc* methodologies, whose development is probably more time consuming than the design itself, have been widely used. In other cases, when the target architecture is dictated by the vendor board, the design is generally well supported by legacy tools but, in this way, the key factor is the experience of the designer in the choice of the most suitable solution.

Hence, a different approach to the problem is necessary. A possible solution is provided by extending the classical co-design methodologies. A co-design environment for the multi-processor embedded system development is of critical importance: it enables the designer to optimize the design process and the design itself by increasing productivity and improving the quality of the whole process. This could be obtained through the definition of a framework that exploits the synergism of the hardware and software parts through their concurrent design.

In this way, the designer, following a co-design flow (co-specification, co-analysis, co-estimation, co-verification/validation, and design space exploration), is able to

- perform as late as possible technology-dependent choices;
- perform easily system-level debugging and testing;
- explore different design alternatives in the architectural design space to identify a solution suitable to maximize system performance and to reduce costs; and
- reduce the system design time by avoiding expensive design loops once arrived at last steps of the design flow.

The framework should allow user interaction to exploit the designer experience at system level, where it is still possible to manage the application complexity.

The next section presents the main issues related to hardware/software co-design, showing its advantages and discussing the implications of a multi-processor embedded system approach. Next, we briefly review meaningful examples of co-design environments that have been developed to support multi-processor embedded system design.

2.2 Concurrent HW/SW Design

Heterogeneous systems are more complex to design than homogeneous ones, because the number of parameters and design choices to be taken into account is much larger than that in the case of a fixed target architecture. Therefore, systematic design techniques are needed, and this has been the topic of research in the area of *hardware/software co-design.*

The need to specify, analyze, verify, and synthesize mixed hardware/software embedded systems at a high abstraction level has been recognized in many application fields such as multimedia [5], automotive [6], mechatronics [7], and telecommunication [8, 9]. A key point in such an activity is the

```
┌─────────────────────────────────────────────┐
│ High-Level Flow                               │
│   • Co-Specification                          │
│   • Co-Analysis                               │
│   • Co-Verification (Co-Validation)           │
│   • Design Space Exploration                  │
│         o  Partitioning                       │
│         o  Architecture Selection             │
│         o  Co-Simulation                      │
│                                               │
│ Low-level flow                                │
│   • Low-Level Synthesis                       │
│         o  Hardware Synthesis                 │
│         o  Compilation                        │
│   • Integration                               │
│   • Low-Level Co-Verification                 │
└─────────────────────────────────────────────┘
```

Figure 2.1   Modern co-design framework.

possibility to take most of the decisions at the system level during the earlier stages of the design, in order to avoid as much as possible design loops including time-consuming synthesis activities.

This has lead to the sharp separation between the high-level and low-level phases of the co-design flow, where the high-level phase represents a unified view of the problem that has replaced the typical separate design approach between the hardware and software parts. The high-level co-design framework is the ideal platform where the designer validates the system functionality and evaluates different tradeoff alternatives before proceeding with the low-level phase of the design, where automated tools perform the synthesis and the integration between the parts is performed before the final low-level co-verification.

A modern co-design framework can be decomposed into several steps (Figure 2.1).

Co-specification

The requirements are translated from an informal language into a formal (or semi-formal) description of the functionalities. An abstract homogeneous behavioral description is given for the complete heterogeneous system, regardless of the target architecture that will be chosen and how the different parts will later be implemented. Using this technology-independent representation, different implementation alternatives can be evaluated before making any commitments.

Co-analysis

Analysis techniques are provided which allow early estimations of the final implementation characteristics (e.g., performance, power consumption, etc.), based on a high-level behavioral description. Such analysis methods are necessary to allow a comparison between different implementation candidates.

Co-verification (Co-validation)

The functional correctness of the system is verified: the specification is simulated to check its behavior with respect to representative test-benches. If the specification is a formal one, it is also possible to prove (in the mathematical sense) its correctness. The main goal is to early identify errors and problems (e.g., deadlock, dead code, etc.) in the specification.

Design space exploration

This phase may be decomposed (at least logically) in three interacting tasks: *partitioning, architecture selection,* and *co-simulation.*

When deciding on the implementation, the designer needs to choose the components to include and how these should be connected in the hardware architecture. It must also be decided which parts of the behavior should be implemented on which of the selected components. The first of these activities is called architecture selection and the second is known as partitioning. Architecture selection and partitioning are influenced by performance requirements, implementation cost, reconfigurability, and application-specific issues. Co-simulation evaluates the system behavior from a functional point of view or a timing point of view, in order to validate either the specification or the performed partitioning.

Low-level synthesis

The partitioned hardware and software specifications are translated into their final form, typically a technology *netlist* for the hardware and an assembly code for the software. The process of translation is referred to as *hardware synthesis* for the hardware blocks and *compilation* for software components. The term synthesis is also used for the software when the behavior of the components is modeled by means of formalisms much more abstract than usual high-level programming languages. Hardware synthesis may require more than one translation step. In such cases, the model, provided by means of abstract behavioral languages, is first translated into an *RT-level VHDL* (or *Verilog*) representation and then fed into commercial *HDL* synthesis tools.

Software compilation, similarly, requires a two-step elaboration. In the first phase, referred to as the front-end, the source code is transformed in an equivalent representation at a lower level of abstraction. At this stage, the functionality is captured by means of complex formal languages such as *p-code*, *byte-code*, *three-operands code*, or more frequently, *register-transfer language*.

The intermediate model is then translated into the target assembly language based on *pattern-matching* and *graph-covering* algorithms. These tasks are usually performed using third-party commercial tools.

Integration

Complex systems often require a significant amount of communication between hardware and software components, thus requiring efficient methodologies for their integration. Such integration (called also interface synthesis) is a critical issue in the synthesis of heterogeneous systems and, in fact, it is often performed manually or only partly automated. In recent years, the use of *Intellectual Properties* (*IPs*) in complex designs is growing rapidly and it is particularly appealing for the implementation of standard interfaces and buses.

Low-level co-verification

The low-level models are simulated with a higher level of detail. At this stage, area, time, and power figures are known and can be used to derive the exact characteristics of the complete system.

At this point, it is interesting to briefly compare how the co-design flow influences the embedded system design process. Figure 2.2 compares a traditional top-down design flow with that of co-design revealing the following major advantages of a co-design approach.

- A detailed high-level specification of the system behavior is made prior to architecture selection and partitioning. This means that, by means of analysis and estimations, more information is available, and these crucial steps can thus be made with increased accuracy.

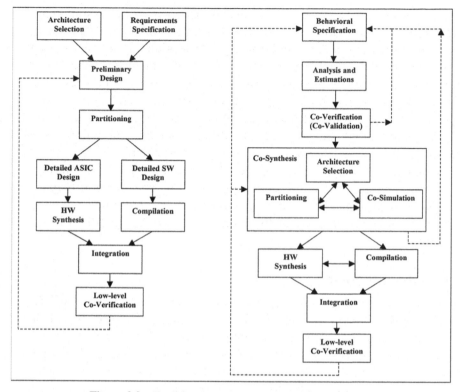

Figure 2.2 Traditional top-down flow vs. co-design flow.

- A uniform description of hardware and software modules allows the exchange of parts of the system between different partitions at later stages of development. It also enhances the possibility of early debugging and validation of the complete system.
- Co-design moves the hardware and software development closer together, thus reducing the cost of the final integration between the different technology domains.

These improvements to the design process allow reduction of development cost and time, resulting in an implementation of higher quality.

However, co-design of heterogeneous multi-processor embedded systems is a more challenging task than classical embedded system co-design since the large number of tuning parameters (e.g., inter-processor communication cost, load balancing issues, etc.) increases the complexity and efficiency of the analysis of the solutions space and adds several degrees of freedom in the *architectural selection* phase. The effectiveness of the design space exploration, i.e., the mapping of the functionalities composing the system specification onto a suitable architecture while minimizing a specific cost function and meeting design requirements, is strongly influenced by the *coarse grain* decisions taken during the early stages of the design process (e.g., number and type of processors, communication strategies, etc.).

A co-design environment, to cope with the task of selecting the most suitable system architecture, should provide a framework where the timing behavior of different configurations of the system is quickly evaluated in a unified manner not only focusing on a microprocessor at a time, but also considering the communication among pieces of the specification mapped on different subsystems. Different *multi-processor oriented* co-design environments have been proposed in the literature, as briefly described in the following.

2.2.1 State-of-the-Art

The interest in co-design research has been steadily increasing from the beginning of the 1990s. The first workshops dedicated to co-design were held in 1991 (in conjunction with the 13th International Conference on Software Engineering, Austin, Texas) and 1992 (International Workshop on Hardware/ Software Co-design, Estes Park, Colorado). The dominating research issue was the partitioning of a behavioral description into an ASIC part and a software part to be executed on a fixed, tightly coupled, processor.

Many co-design environments have been developed from then: they were principally devoted to 1-processor/n-ASIC architectures. Some of the most representative ones have been: *POLIS* [73], *COSYMA* [113], *COMET* [114], *COOL* [115], *COSMOS* [116], *Chinook* [117], and others.

Some of these were already oriented (at least in principle) to multi-processor or distributed target architectures. A representative example is *CoWare* [74, 118], a co-design environment today commercially available. It was developed to support the design of distributed embedded systems and, nowadays, it is targeted to mono/multi-processor systems-on-chip.

In the past few years, an increasing number of research works has considered the problem of defining a co-design methodology for heterogeneous multi-processor embedded systems. The analysis of the literature has allowed the identification of the main benefits and drawbacks of the proposed approaches, thus identifying the most critical issues to be considered in the present book.

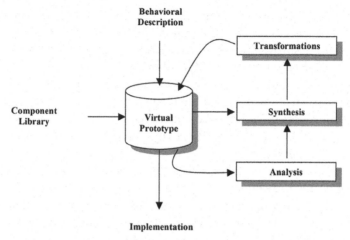

Figure 2.3 Virtual prototype based framework.

Analysis and synthesis of heterogeneous real-time systems

The considered work [3, 142] proposes a consistent framework for real-time system design, integrating analysis models and synthesis tools in an effective way. The work presents a hardware/software co-design approach for the development of real-time systems, looking for an implementation that satisfies the specified timing constraints. The target architecture is heterogeneous based on a mixture of microprocessors and application-specific integrated circuits. The design space exploration is performed at the system-level using semi-automatic synthesis tools, which operate on virtual prototypes of the implementation. The design flow is shown in Figure 2.3.

The behavior of the entire system is specified in a high-level, homogeneous description, independently on how different parts will later be implemented. The book proposes a semi-formal abstract behavioral model (i.e., it is not targeted to any common specification language) that however is not executable. The lack of this property is a serious one for systems where functional correctness and timing analysis cannot be performed in a static manner.

The target architecture is a heterogeneous multi-processor assembled from a component library that includes processors, *ASICs*, and memories connected by means of a single bus.

The aim of the synthesis phase is to minimize the implementation cost, while meeting all timing constraints. The work presents also an *intrinsic analysis* methodology, which estimates the hardware resource usage of individual tasks when considered in isolation on the architecture, and an *extrinsic analysis* methodology for determining the effects of resource sharing between several concurrent tasks. A general estimation model is proposed to analyze the characteristics of the software parts of a system at a high level of abstraction, and to provide information synthesis algorithms.

The intrinsic analyses are performed by heavily exploiting the limitation imposed on the specification (e.g., bounded loops, i.e., not an atypical assumption in the real-time world). Moreover, estimation models have been developed only for *CISC* processors and the *ASICs* analysis has been based only on the work present in the literature.

The extrinsic analysis is based on a schedulability model (used also during the co-validation step) to evaluate the so-called *minimal required speedup*, i.e., how much the underlying hardware would at least have to be accelerated to satisfy the requirements. The only drawback of this technique is that it requires information about the tasks that are rarely available while designing general systems at a high-level of abstraction. It is instead suitable for well-known real-time systems. The subtasks of the co-synthesis step are: task priority assignment, partitioning, and architecture selection.

The first step is performed considering different models for single-processor and heterogeneous systems with independent tasks, adding in the last step heuristics devoted to consider inter-task communications. The partitioning subtask is based on a *branch-and-bound* approach and its role is limited to select the appropriate executors for each system task.

The final co-synthesis subtask is performed by the architecture selection step that, based on the results provided by the previous subtasks, defines the details of the final system. To this purpose, a quality function used to compare several solutions is properly defined, and many heuristics (e.g., genetic algorithms, taboo search, simulated annealing, etc.) are compared with extreme detail. However, there is a lack of analysis (and validation) on the overall quality of the results that such heuristics are able to provide.

The co-validation of the system correctness is performed statically by means of a fixed-priority scheduling that introduces a total ordering on the tasks set. The duration of each task is determined by means of bounded loops and critical path analysis (already performed during the co-analysis step). Hence, it is possible to analyze the response time of the system independently of the input data sets. However, such an approach is feasible only for particular categories of systems. If the proposed solution does not meet some of the constraints, some transformations can be applied to the scheduling, the partitioning, or the architecture.

This research work represents surely a step toward a more engineering-like system-level design practice for heterogeneously real-time systems. The author argues that the use of the presented techniques should increase the predictability of the design process since it allows performance problems to be discovered and eliminated early during the design. It should also shorten the time-to-market and reduce the development cost, because many design steps can be automated (at least partially). These claims are justified through theoretical results and practical experiments.

In conclusion, this work could be considered as a reference one, where the hypothesis and the assumptions related to its application domain (i.e., real-time systems) should be replaced with different ones that could lead to an exacerbation of some aspects (e.g., time analysis) but probably allowing the definition of a less application-domain-oriented system-level co-design environment.

COSYN and COHRA: co-synthesis of (hierarchical) heterogeneous distributed embedded systems

COSYN [119] presents a co-synthesis algorithm that starts with an embedded-system specification and results in an architecture consisting of hardware and software modules to meet performance, power, and cost goals.

The specification is provided by means of a set of annotated acyclic periodic task graphs. This choice differentiates this approach from the one presented in this book. In fact, such a specification formalism is suitable only for a restricted range of application, and granularities issues prevent from taking a real system-level approach for complex systems (the authors show the application of the algorithm to a real-life example but the results, especially the number of processing elements, could give rise to a certain form of skepticism). In general, this is true for all *task graph-based* co-design environments.

The target architecture is detailed by means of *resource libraries* and by means of several description vectors (*execution_vector*, *preference_vector*, *exclusion_vector*, etc.). Such vectors provide every useful detail about the system behavior and characteristics (e.g., timing).

The co-synthesis algorithm is composed of several steps: association, clustering, allocation, scheduling, performance estimation, and power optimization. It employs a combination of preemptive and non-preemptive static scheduling. It allows task graphs in which different tasks have different deadlines. It introduces the concept of an association array to tackle the problem of multi-rate systems. It uses an interesting task clustering technique, which takes the changing nature of the critical path in the task graph into account.

COHRA [120] is an extension to the previous one made by the same authors. They provide to the first approach the capability to manage hierarchical task graphs, improving the quality of the results, in particular for large task graphs. However, the same considerations made for the first work apply as well to this second one.

These approaches are very interesting and detailed and it seems to provide excellent results on small examples, however, each step of the algorithm is strictly related to the task graph theory and it seems very difficult to express general systems by means of such a specification formalism. Such a difficulty imposes limitations on the possibilities of integrating the proposed co-synthesis methodology in a more general co-design environment.

S^3E^2S: Specification, Simulation, and Synthesis of Embedded Electronic Systems

The work presented in [96] describes the system synthesis techniques available in S^3E^2S, a *CAD* environment for the specification, simulation, and synthesis of embedded electronic systems that can be modeled as a combination of analog parts, digital hardware, and software.

S^3E^2S is based on a distributed, object-oriented system model, where objects are initially modeled by their abstract behavior and may be later refined into digital or analog hardware and software. It is built on top of SIMOO, an integrated environment for object-oriented modeling and simulation of discrete systems. SIMOO is composed of a class library and a model editor.

In spite of the fact that the S^3E^2S modeling environment could manage heterogeneous system, the considered target architecture focuses only on a multi-processor paradigm that does not consider *ASICs* or *FPGAs*. This is based on a library of processors, each with different characteristics, ranging from microcontrollers to digital signal processors, with different architectures available in each domain. Each object of the specification may be mapped to a single processor, and each processor may execute the function of one or more objects.

The evaluation of software performance is based on a two-step procedure. First, a high-level processor-independent representation is obtained, like a *CDFG* (control and data flow graph), and then it is translated into a sequence of machine-independent 3-address codes. In order to better analyze each processor, three types of virtual machines were defined, according to different target architectures, such as microcontroller, *RISC*, and *DSP*. The next step concerns object analysis, where the dominant characteristic of the object is identified: control intensive (many control instructions and flow breaks), memory intensive (list processing, digital filtering, heavy memory use), or data processing intensive (few memory accesses, most processing done on internal registers). Each one of these characteristics favors a different processor in the library.

The major drawback of this approach is that it is a *back annotation-like* one, that forces a low-level analysis in order to make choices at the system level.

The last step is related to co-synthesis that involves mainly the allocation of the objects on the associated processor, and the serialization of the functions, that is, to try to group a set of objects on a single processor in order to reduce the cost. At the beginning of this allocation process, all actions that the user requires to be executed in parallel will be necessarily allocated to different processors. Regarding other actions, the communication protocol is checked. In case of synchronous communication, actions of the communicating objects are naturally sequential and may be allocated to the same processor.

The main drawback of the allocation process is the granularity. It works with a too coarse granularity level where whole objects are considered instead of single methods, leading to the allocation on the same processor of functionalities that could be very different.

In conclusion, the whole work presents innovative and interesting ideas but many aspects of the methodologies adopted in the various steps could be improved.

CMAPS: A Co-synthesis Methodology for Application-oriented General-purpose Parallel Systems

The work presented in [121] tries to shift the effort of the research in the co-design field from system design to requirements analysis. The main issue is to analyze how an application problem is transformed into specifications and to exploit this knowledge to improve the whole co-design process. The goal is to design a system starting from the application problem itself, rather than from a detailed behavioral specification.

The target architectures are the so-called *application-oriented general-purpose parallel* (*AOGPP*) systems, which are defined as general-purpose systems with their subsystems designed for the efficient execution of some software solution to a given problem. However, such a definition is all but clear.

A user can specify a complex application problem by referring to the *elementary problems* in a *Problem Base* and describing how the selected elementary problems compose into the desired application problem. Upgradability is made easy using *elementary algorithms*, which act as off-the-shelf building blocks for software and by the use of subsystem architecture models for hardware. Three repositories are used in such a methodology, namely *Problem Base* (*PB*), *Algorithm Base* (*AB*), and *Model Base* (*MB*), which represent the modularizations of specification input, of software synthesis, and of hardware synthesis, respectively.

PB is used to store elementary problems and related information such as the unique problem name and pointers to the corresponding elementary algorithms that can be used to solve the specific problem (e.g., sorting a sequence, solving a set of linear equations, computing the discrete Fourier transform, etc.). Naturally, the class of problems that can be solved is seriously limited by the problem base nature.

AB is a collection of elementary parallel algorithms that can be used to solve the problem in PB. Related information, such as the time and space complexities, and the requirement restrictions on the hardware architecture are all stored along with each algorithm.

MB is a repository of models for hardware subsystems, such as *Communication Modules* (*CM*), *Memory Latency* models (*ML*), *Memory Access* models (*MA*), and *Control* models (*CO*).

Designers can input their specifications by constructing a *Problem Graph* using elementary sub-problems from the PB, along with the sub-problem size and other related constraints. First, CMAPS maps this graph into a *Solution Graph*. Then, CMAPS transforms such initial solution into hardware and software models: the Solution Graph is made feasible iteratively through an interleaving of hardware and software modeling process.

Then a co-evaluation phase is performed to reduce the number of hardware and software modules to be considered in the synthesis phase, thus decreasing the complexity of the co-synthesis step.

Finally, the hardware and software models are synthesized into hardware system-level specifications and parallel pseudo-programs, respectively, and a co-simulation of hardware and software is performed by choosing an appropriate scheduling algorithm.

The methodology presented in such a work is a very interesting attempt to raise the abstraction level of the entry-point of a co-design environment. However, it is far from a widespread applicability. In fact, as enforced by the examples provided in the work itself, it seems very difficult to *compose* an application different from those related to some standard functionality (e.g., to solve linear equations, sorting, matrix transposition, discrete Fourier transform, etc.).

Other works

Other than the environments presented above, other approaches consider the same problem from different points of view.

A first set of co-design environments, targeted to multi-processor systems, relies on the use of commercial tools. For example, in [30], a Simulink-based [122] approach to system-level design and architecture selection is presented. It is suitable for specification and analysis of data-dominated embedded systems and generates *VHDL* and *C* for hardware and software modules, respectively. However, for specification, simulation, and performance estimation, it relies on a library of modules that should be properly composed in order to describe the system. The target application area is restricted to the real-time control. Another example is *MAGIC* [123], a co-design environment of embedded real-time multi-processor signal-processing systems based on Matlab [122] and Excel [124]. It allows the designer to capture the specification in an executable model that can be used in design exploration to find the optimal *COTS* technology and architecture. However, the environment is limited to system-on-board designs (the system supports the products of some of the major vendors [110, 112]).

Another set of research works, related to co-design of multi-processor embedded systems, approaches the problem at a level lower than the system one. For example, [125] presents a co-design flow for the generation of application-specific multi-processor architectures. In the flow, architectural parameters are extracted from a high-level system specification and they are used to directly instantiate architectural components, such as processors, coprocessors, memory modules, and communication networks. However, the micro-architectural level adopted in such an approach limits the design space exploration: it could be interesting to integrate this environment with a system-level-oriented one.

2.3 Conclusion

This chapter has presented the main concepts related to heterogeneous multi-processor embedded systems, describing their general architecture, their applications, and showing the complexity of their design. With the purpose of describing a possible design approach for such systems, the co-design approach and its main issues have been introduced, showing the advantages with respect to classical approaches and discussing the implications of a heterogeneous multi-processor approach. Moreover, the chapter has presented a review of the co-design environments that have been recently developed to support multi-processor embedded system design, identifying the general drawbacks that should be overcome.

What is still needed is a systematic approach to the co-design of such systems oriented toward the system-level, quite general to be useful in several application domains, based on assumptions that do not limit its applicability, but able to extract and properly consider the relevant features of the system to be designed.

3

The Proposed Approach

In the field of heterogeneous multi-processor embedded systems design, a co-design framework is of critical importance: it enables the designer to optimize the design process and the design itself. This work is an attempt to fulfill the top-level requirements of a co-design environment able to support the concurrent design of embedded systems, possibly subject to real-time constraints, implemented on heterogeneous multi-processor architectures.

The starting point of this work is the *TOols for System Co-Design Automation* (*TOSCA*) environment [4, 9, 11, 62], an existing single-processor-based co-design environment. Such an environment has been extended to support, at system-level, co-specification, co-analysis, co-simulation, and design space exploration of heterogeneous multi-processor embedded systems, providing directives to guide low-level tools (e.g., compilers hardware synthesizers, etc.). Therefore, before the detailed analysis of the proposed co-design flow, Section 3.1 provides an overview of such reference environment, and then Section 3.2 introduces the proposed approach.

3.1 The Reference Environment: TOSCA

The hardware/software co-design environment *TOSCA* has been developed to study and to propose a solution to some of the many problems that arise when designing heterogeneous systems using both hardware components (*ASICs*, *FPGAs*), *off-the-shelf* cores (*microprocessors*, *microcontrollers*, *DSPs*), and software programs.

Figure 3.1 shows a coarse-grained outline of the co-design flow, where the grayed-out boxes indicate the portions belonging to the TOSCA environment. The portion outside these boxes represents third-party commercial tools.

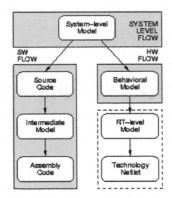

Figure 3.1   Outline of the TOSCA co-design flow.

A common portion, at the beginning, deals with the functional model of the system, neglecting all the partitioning and implementation issues. The flow then splits into two paths: one for the software components and one for the hardware components. In each path, the highest levels are completely technology independent while the lower levels constitute the actual implementation of the system. The intermediate levels have the twofold purpose of providing a more detailed description of the functionalities modeled at a high level while still maintaining an abstraction sufficient to neglect the technological details. In this section, the focus is on the low-level flow while, in the next one, the multi-processor-oriented system-level flow is outlined.

3.1.1 The Specification Language

The *TOSCA* environment follows a single language approach in which a homogeneous specification of the system functionalities is given by means of the *OCCAM* language (namely OCCAM 2). OCCAM [12] is a programming language developed by *INMOS* in the 1980s to exploit the parallelism of *transputers*. The OCCAM language offers a number of complex constructs and features that are currently not supported by TOSCA. In particular, the last version of OCCAM, called OCCAM 3, supports dynamic memory allocation and pointers. The supported subset for embedded system specification in TOSCA is summarized in Table 3.1.

This language offers some features that are extremely attracting for mixed hardware/software specifications and it is described with more detail in Section 4.2.1.

Table 3.1 OCCAM subset supported by TOSCA

Construct Class	Supported
Container	*PROC, PAR, SEQ*
Control	*ALT, IF, WHILE*
Basic Processes	Assignments, *SKIP, STOP*
Communication	! (output), ? (input)
Variables	Channels, Variables, Constants
Types	*BIT, BYTE, INT*

3.1.2 Intermediate Representations

Within the *TOSCA* flow, the functional high-level model undergoes a number of subsequent transformations that lead to the final technological representation adherent to the characteristics of the target architecture described in the next paragraph.

Different languages and formalisms have been used to describe the intermediate stages, some being standard or well-known, other being defined from scratch or being customizations of popular formalisms.

The most notable language introduced in TOSCA, namely in the TOSCA software flow, is called *VIS* or *Virtual Instruction Set*. This language closely resembles an actual assembly language (in particular the *Motorola MC68000* assembly language [13]) but at the same time differs from it to some extent.

The VIS language plays a fundamental role in the software flow: on one hand, it allows for easy re-targeting and, on the other hand, it introduces a new, almost technology-independent level of abstraction at which analysis can be carried out.

The most relevant characteristics of the VIS language are summarized in the following points.

- *Target processor independence*: the VIS provides the most common operations of any assembly language but remains sufficiently general to allow easy translation (*mapping*, in the following) into most of the actual assembly languages.
- *Customizability*: the language is flexible with respect to some parameters, in particular the number of general purpose registers available is not fixed but can be defined at compile time. Similarly, the stack growth direction and the alignment requirements are user-definable.
- *Simulatability*: the VIS code can be simulated using the VIS simulator (debugger and profiler) that has been developed within TOSCA.

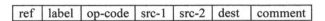

| ref | label | op-code | src-1 | src-2 | dest | comment |

Figure 3.2   The VIS language syntax.

- *Modularity*: thanks to a certain number of directives, the VIS language allows the development of different modules separately. The modules can be eventually linked into a single complete, stand-alone, simulatable code. To this purpose, a collection of basic functions, the *VIS Operating System*, has been developed. These functions provide memory allocation facilities as well as communication primitives and the kernel for parallelism emulation.
- *Cross-level referencing*: the presence of a special *source code reference* field and two directives allows the language to maintain references to the source language from which it has been originated.

VIS is a three-operand assembly-like language whose syntax is summarized in Figure 3.2. The general structure of the language is similar to that of a *RISC* machine since it refers to load/store operations.

3.1.3 The Target Architecture

The transformation of a functional, high-level specification into a working system made of *ICs*, buses, memories, microprocessor cores, etc., is an extremely complex task and involves an enormous number of choices. The problem has thus a high number of degrees of freedom and it is not realistic to expect a tool to be capable of undertaking all the choices automatically. For this reason, a certain degree of interactivity with the designer is necessary and some *a priori* assumptions have to be made on the target architecture.

Within the *TOSCA* environment, the following assumptions are made:

- *Hardware*: the target architecture can have up to 256 hardware functional units. Each unit has input and output *FIFOs* and status registers storing information on the status of the FIFOs. Each unit is called *coprocessor* and it corresponds to an *OCCAM PROC*.
- *Software*: the software portion of the system runs on a single microprocessor core.
- *Bus*: the main communication medium is a single bus shared by the processor and the coprocessors.

The shared bus provides a physical medium to interconnect all the functional elements of the architecture. The way the bus is used depends on the type of channel communication involved, specifically:

- *Software/Software*: the channel is managed using a data structure in a global memory area; all procedures can access the global memory area simply addressing it properly.
- *Hardware/Software* and *Software/Hardware*: the FIFOs of the coprocessors are memory mapped and thus the software portion of the design has no knowledge of the different nature of the target of a channel. Owing to the different typical speed of the hardware and the software partitions, input/output is buffered and it is assumed that the coprocessors are fast enough to consume data present in the input FIFOs quickly enough to avoid overflows. For the same reason, the output FIFOs tend to fill-up quickly: when such a situation occurs, a suitable interrupt signal is raised and the coprocessor is disabled until some data are read from its output FIFOs. The status register mimics the data structure used to store channels in memory.
- *Hardware/Hardware*: unless explicitly specified during the hardware compilation phase, the bus is not used for this kind of communication and dedicated point-to-point connections are used instead.

The target architecture is sketched in Figure 3.3 and the internal structure of a coprocessor is depicted in Figure 3.4.

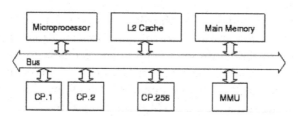

Figure 3.3   The TOSCA target architecture.

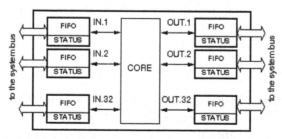

Figure 3.4   The TOSCA co-processor internal structure.

3.1.4 Overview of the Design Flow

The *TOSCA* framework is based on the use of a modified version of the *OCCAM-2* language as input formalism for the description of the entire hardware/software system. This description can be input in text mode and in a mixed graphical/textual mode. Currently, the graphical language provides the capability of describing parallelism and communication only, while the algorithmic portion must be entered in the text format.

The system-level OCCAM model is the starting point of both the hardware and software flows. This model can be simulated in different *modes*, namely:

- *Functional Mode*: only the system functionalities are simulated. Timing aspects are neglected in the sense that all the operations and the transitions are considered instantaneous. This mode does not distinguish among hardware, software, and test-bench partitions.
- *Partitioned Mode*: different partitions are characterized with different operation and transition times. These times may be either actual, i.e., derived from technological libraries, or fictitious.

At this very high level of abstraction, it is thus possible not only to verify the functional behavior of the model but also to derive some preliminary estimates of its timing characteristics.

After this first verification step, the flow is divided into two parallel branches: the *hardware flow* and the *software flow*. Some portions of the design model are assigned to the hardware partition, some others to the software partition, and those modeling the test-benches are left unassigned. These portions are then elaborated following one of the two flows. The granularity chosen for partitioning is that of OCCAM procedures *PROC*.

The hardware flow is divided in a *front-end* elaboration, which is part of the TOSCA framework, and a *back-end* processing, based on commercial *VHDL-based* tools [106]. The front-end is constituted by a behavioral synthesis tool that translates the OCCAM model into a VHDL description at the register transfer level. The synthesis algorithm is based on some assumptions in order to limit the complexity of the problem.

The hardware flow

The core of the hardware flow consists in a translation of the OCCAM system-level model into a VHDL description at RTL level. Such a translation is similar to a behavioral synthesis since the OCCAM model does not contain explicit clock signals and does not specify the resources available.

The problem of translating an algorithmic, high-level description into a register-transfer model is rather complex, regardless of the description languages used. For this reason, the TOSCA *OCCAM-to-VHDL synthesizer* is based on some simplifying assumptions, the most important ones are reported in the following.

- The synthesis is *template based* in the sense that some OCCAM processes are translated into fixed data-path and control templates.
- An *OCCAM PROC* is translated into a *VHDL ENTITY* with the procedure formal parameters mapped to the input/output *PORTs*.
- The *OCCAM input* (?) and *output* (!) processes are synthesized by instantiating two specific library components. These components implement the semantic of the rendezvous communication protocol.
- The semantic of containers (*PAR*, *SEQ*, etc.) is implemented by using parametric control templates. The templates are constituted by a simple finite-state machine and some combinatorial logic made scalable by means of a combination of *GENERICs* and *GENERATE* statements.
- OCCAM variables are mapped to VHDL SIGNALs.
- *OCCAM channels* are mapped to buses with a predefined structure.

These assumptions allow the architecture of the OCCAM model to be translated into *RTL (Register Transfer Language)* code in a rather straightforward way. Nevertheless, the optimality of such a synthesis is far from being achieved.

Once the control structure has been fixed according to the previous rules, the data-path must be synthesized. The current implementation of the tool performs operations scheduling and binding under resources constraints and operators sharing within the same OCCAM container process. The proposed hierarchical, template-based recursive translation methodology produces correct VHDL but the optimality of the result is far from being achieved.

The software compilation flow

The portion of the design tagged as software needs to undergo a number of transformations that will lead to the complete assembly code.

The starting point is a *TOSCA project*, i.e., a collection of *OCCAM* procedures, with no notion of hierarchy or scope, organized as follows:

The project root directory contains a number of sub-directories, one for each procedure, devoted to contain all the information related to the specific procedure and a project file.

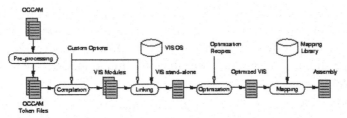

Figure 3.5   The TOSCA compilation flow.

Each sub-directory contains a plain text file called *OCCAM* that is the OCCAM source code. This file can be either handwritten or automatically generated from a graphical model by means of the exporting facility of the *OCCAM graphical editor*. A procedure named *Main* must always exist and constitutes the entry point of the code. The project file, named *project*, is a list of the names of the procedures followed by an indication of the partition assigned to the procedure.

The compilation flow, sketched in Figure 3.5, begins with a translation of the OCCAM source code (either of a single procedure or of the whole project) into a VIS program. This phase is an actual compilation and contains most of the complexity of the entire forward flow. The generated VIS files are then linked, optimized, and finally mapped to the desired assembly language. The rest of this section details the phases of the forward flow.

Pre-processing

Prior to compilation, *OCCAM* files must be preprocessed. This is necessary since the OCCAM language has a syntax that depends on the indentation of the code and thus no *LR parser* can read it. Fortunately, it is rather easy to build a filter that reads the OCCAM code with the proper indentation and generates a more solid parenthesized representation. Such a format is then read by a *bison-generated LR parser* [44] and a *token file* is produced.

The parser, thus, does not directly build a syntax tree in memory but rather generates a simpler description of the contents of the source file. The token file generated can be read with a very simple C++ [24] procedure that only makes use of some stacks to temporarily hold the tokens read.

Compilation

The compilation is driven by user-defined options that determine some of the properties of the *VIS* language being generated and that drive the behavior of the compiler itself. Compilation requires that some of the generality of the VIS language be lost since the following parameters must be supplied:

- *The number of general purpose registers (GPRs) available.* This, of course, influences the generated code, in particular the translation of expressions and conditions and the number of temporary variables allocated.
- *The memory alignment.* Some architectures may only access memory at half-word or at word boundaries. For this reason, the variables, depending on their type, must be allocated properly, i.e., at proper memory location.
- *The stack growth direction.* The *pop* and *push* instructions of an actual target assembly assumes a fixed stack growth direction and thus the memory allocated for the stack at VIS level as well as the stack pointer register (*SP*) must be initialized accordingly.

These parameters affect the generated VIS code in a way that may compromise the semantic adherence to the OCCAM model.

Other user-definable parameters (e.g., the *spilling policy*) alter the behavior of the *OCCAM-to-VIS compiler* with respect to the optimality of the generated code, without any risk for its correctness. These parameters are described in [10].

The compilation process itself is the result of five subsequent processing phases referred to as *compilation passes*. Each pass uses the information gathered in the previous pass and either generates a new representation of the OCCAM model or completes the existing one. The operations performed by the five passes are: *Temporary variables allocation, Symbol table construction, Symbolic code generation, VIS code generation,* and *VIS code instrumentation.*

The instrumentation mechanism is rather peculiar and constitutes one of the main issues in parallel language compilation. More details and some examples are given in [10].

Linking

The result of compilation is a VIS file for each procedures of the project. These files need to be linked into a single *VIS* program.

The linking phase is controlled by some user-specified parameters. Note that these parameters must be compliant with those used for compilation. As an example, the linker must know in advance the number of registers available, and this number must match the number of registers used for compilation.

The linker resolves a number of external references and generates the necessary initialization code.

To enhance efficiency, different versions of the VIS system routines have been written and optimized for the number of available general-purpose registers. The result of linking is a single, *stand-alone*, simulatable program.

Optimization

The *VIS* code of each procedure is slightly modified by the linking process and its optimality is, to some extent, lost. For this reason, a *machine-independent optimizer* has been introduced in the flow. The optimizer reads in the VIS code, breaks it into basic blocks and builds the flow graph on which to operate. The optimization phase is driven by an *optimization recipe*, i.e., a sequence of different optimizations selected from the set of all available optimizations in order to reduce the code size and/or increase execution speed.

Some of the classical optimizations that the tool currently performs are *NOP Elimination, Constant Propagation, Dead-Code Elimination, Branch Reduction*, and *Simple Loop Unrolling*.

Mapping

The *VIS* code generated by the optimizer is eventually translated into a specific target assembly language. This last step is performed by a mapping tool, the *mapper kernel*, with the support of processor-specific mapping libraries. The mapper kernel reads the VIS code, one line at a time, and searches the library for the suitable mapping rule to apply. A mapping rule is a C function determining how each VIS instruction, along with its addressing mode, has to be translated into the specific assembly language. All the rules for a given processor are collected in a mapping library, which is a shared object library file that the kernel can link at run-time. The library contains also three special rules that are used to add a header at the beginning of the generated assembly code, to add footer for housekeeping and to re-map the VIS registers to the actual registers.

Currently, the mapping rules are *one-to-many* rules, and this impacts the quality and the optimality of the generated assembly because coupling of adjacent VIS instructions is neither considered nor exploited.

This step concludes the *forward flow*. Note that, at each step, reference files, containing information on how the source code has been translated across the lower levels, are generated. These files will turn out to be fundamental for the *back-annotation* process.

The estimation flow

The estimation flow implements the methodology described more in detail in the following of this book, for the case of *performance estimation*. Other kinds of estimations, especially the *power consumption* one, are described in [10].

The flow operates at the three levels of abstraction that have been described in the previous section: *target assembly* level, *VIS level,* and *OCCAM level.*

The timing data, along with power consumption and area data, are obtained with three different mechanisms:

- *Annotation*: the code is annotated associating with each line a set of figures obtained from a library. The annotation library collects area, timing, and power figures derived either from direct measurement or from mathematical models built on a statistical basis.
- *Estimation*: the code is annotated with figures calculated using statistical models collected into libraries. The difference between annotation libraries and estimation libraries is that the latter are a collection of higher-level models of the language built on the base of a statistical analysis of the relations between the language under consideration and lower level languages.
- *Back-annotation*: the code is annotated with figures directly derived from lower-level data. This implies that annotation or estimation needs to be performed at least at one of the levels of abstraction below the level under consideration.

The figures obtained with these three processes are *static* in the sense that they do not depend on an actual execution of the code but rather indicate the timing required by each instruction in the code. To obtain a complete estimate, it is necessary to include in the characterization the profiling information derived from the *functional* simulations of the code with typical input data. The profiling data are combined with static figures to give a *dynamic* estimate.

Moreover, the static estimates for each OCCAM process are fundamental to perform high-level *timing* simulations in order to evaluate delays and latencies due to communication and synchronization between the systems components.

Annotation and estimation

From an abstract point of view, the methodologies for annotation and for estimation are identical: they both read a source file in one of the languages used in the design flow, search some data in an external target-dependent library, and annotate each line of the source code with static figures.

The main differences are in the way the libraries have been derived and the models on which they back up. These differences have been analyzed in [10].

Another difference is related to the language being read: the *OCCAM* and the *VIS* languages are independent of the target processor while the assembly code is determined by the choice of the processor. Therefore, the parsing capabilities are embedded in the core of the tools for the first two cases while they are moved to the libraries for the last case.

This choice isolates the target-processor-dependent knowledge into the libraries, making the tools independent. For this reason, the assembly annotation libraries cannot be a collection of data only but rather a collection of functions embedding the power, timing, and area figures. The assembly annotation tool thus dynamically links the library and obtains access to the desired functions.

Back-annotation

The back-annotation mechanism allows reporting figures, derived at the lower levels of abstraction, up in the flow to the more abstract descriptions. In general, a single line of *OCCAM* code is translated in 5−20 *VIS* instructions, which in turn correspond to 10−40 assembly instructions.

This means that when reporting information toward the higher levels, data must be suitably collected. To do this, it is necessary to have access to the reference files generated in the forward flow. In particular, a VIS reference file contains, for each VIS line, a reference to the OCCAM line and to the file from which it has been originated. Similarly, at assembly level, for each line of code the reference file contains an index to the corresponding VIS line. Figure 3.6 shows a sample code at the three levels of abstraction and the corresponding reference file and clarifies the referencing mechanism.

The next section provides an overview of the extensions made to the described reference environment in order to support heterogeneous multiprocessor embedded systems design, focusing on the system-level flow (Figure 3.1).

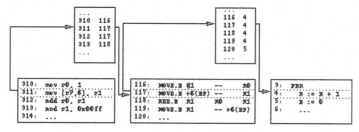

Figure 3.6 Back-annotation and cross-references.

3.2 The Proposed Environment: TOHSCA

The extension to the TOSCA environment has been made following two primary goals: allow the environment to fully support at system level the heterogeneous multi-processor embedded systems co-design process (specification, analysis, estimation, simulation, and design space exploration) and integrate the new high-level flow with the existing low-level one. The proposed approach is then characterized by the following main features:

- homogeneous behavioral system-level specification representing the system functionality and the timing constraints;
- analysis of the specification in order to statically detect the best processing element for each system functionality, and to statically estimate their timing characterization for both hw and sw implementations;
- system-level functional co-simulation to check the functional correctness of the specification and to provide a set of dynamical information on the system behavior (profiling, communication, load, etc.); and
- system-level design space exploration composed of two integrated and iterative steps:
  - a partitioning methodology exploring the design space to identify feasible solutions, supporting heterogeneous multi-processor *architecture selection* (number and kind of heterogeneous processing elements) taking into account several issues (degree of affinity, communication cost, processing elements load, concurrency, area, physical cost, etc.) and
  - system-level timing co-simulation in considering heterogeneous multi-processor architectures and a high-level model for the communication media, to check the satisfaction of the timing constraints.

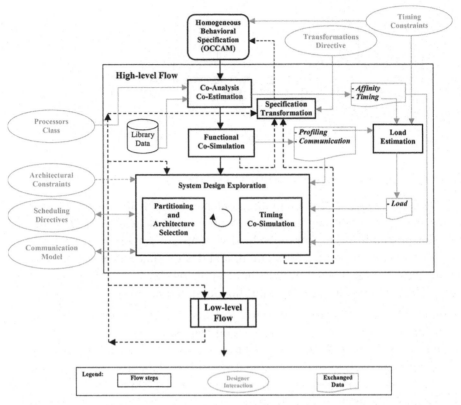

Figure 3.7   The proposed high-level flow.

The last part of this chapter briefly analyzes each step of the new co-design flow (providing also a detailed description of the book structure) and reviews the relevant features of the new target architecture emphasizing the difference with the previous one.

The Proposed Flow

The new high-level flow can logically replace the system-level flow of TOSCA (Figure 3.1): this allows the old environment to be extended by keeping the integration with the existent low-level flow. The main implication of such integration is the possibility of reusing the existing tools and to interface with the same commercial products, thus obtaining an economic and *ready-to-use* back-end. In the rest of this book, in order to explicitly show the common parts between the old and the new environment, TOHSCA (*TOols*

for Heterogeneous multi-processor embedded Systems Co-design Automation)
is used to refer to innovative issues, while TO(H)SCA indicates a common
feature for both the environments.

Figure 3.7 shows a detailed description of the proposed high-level flow.
The main steps are identified with the used and produced data, and the possible
interactions with the designer. In the following, each step is described in detail.

Co-specification

The entry point of the proposed co-design flow is the OCCAM specification
of the desired system behavior. The use of such a homogeneous language
allows the designer to ignore the implementation details in order to avoid
polarizing the design toward hardware or software solution at this early stage of
the flow. Moreover, thanks to some extensions made to the OCCAM language
(described in Chapter 4), the designer can specify various kinds of timing
constraints on the specification: the satisfaction of such constraints is the
primary design goal.

Chapter 4 analyzes system design issues and describes the main system-
level specification languages, focusing on OCCAM as the reference language
and on System C as a promising standard. Moreover, Chapter 4 describes the
internal models developed to represent the OCCAM specification.

Co-analysis and co-estimation

The first step of the flow aims at obtaining as much information as possible
on the system, analyzing the specification in a static (and fast) manner. More
in detail, the goals of this step are twofold: for each system functionality,
to statically detect the best processing element for their execution, and to
estimate their timing characterization for both HW and SW implementations.
This step provides a set of data expressing the *affinity* of each functionality
toward each possible processing element (GPP, DSP, ASIC/FPGA), and a set of
estimations on the *time* needed, to a particular class of processing elements,
for the execution of each single operation that composes the specification.
The timing estimation tool requires, for the software case, the specification of
the processors class that will execute the code (i.e., x86, SPARC, MC680x0,
ARM, etc.), and a data library associated with the selected class (obtainable
from processor documentation) and to the possible hardware technologies. It
is worth noting that the estimation tool is dependent on the processors class
only through such library, and so, it can be used potentially for every kind of
processor.

The model, the methodology, and the tool related to the co-analysis are described in Chapter 5, while Chapter 6 discusses the co-estimation issues.

Functional co-simulation

After the static analysis, the system functionalities are simulated in order to verify their correctness with respect to typical input data sets. This kind of simulation is approximated but very fast and allows the designer to easily detect functional errors. In fact, the timing aspects are not considered but other issues, such as synchronization, precedence, and mutual exclusion between OCCAM processes can be observed, eventually detecting anomalous situations such as deadlocks or the presence of dead code. Moreover, it is possible to extract important data characterizing the dynamic behavior of the system: *profiling* and *communication cost*.

This means that it is possible to evaluate the number of executions of each OCCAM process, the amount of data exchanged between processes, and the set of procedures that typically run concurrently on the system (these information are always related to the behavior of the system in correspondence of typical input data-sets). Finally, the early detection of anomalous behavior allows the designer to correct the specification avoiding a late discovery of problems that could lead to time-consuming (i.e., costly) design loops.

The functional co-simulation is performed by the same tool that performs the timing ones. So, all the co-simulation issues are analyzed in Chapter 8 focusing more in detail on the timing analysis.

Load estimation

Combining some of the data provided by previous steps (timing and profiling data) with the designer imposed timing constraints allows the estimation of the *load* that each OCCAM procedure imposes to a processor (*GPP*) that executes it. The extraction of these data from a behavioral specification is an important task that allows for, during the system design exploration step, the evaluation of the number of needed processors and the identification of those procedures that probably need an executor more performing than a GPP. The load evaluation is still very important because it provides information that are rarely available in co-design flows starting from behavioral specifications.

The load evaluation task is strictly related to the functional co-simulation results, so it is analyzed in the same chapter (i.e., Chapter 8).

System design exploration

In this flow, the system design exploration is constituted by two iterative steps: *partitioning and architecture selection*, and *timing co-simulation*. All the data produced in the previous steps are used to guide the process, together with additional information provided by the designer. Such information expresses the *architectural constraints* (e.g., max number of GPP, max number of DSP, area limitation for ASIC, etc.), the *scheduling directives* (e.g., procedures priority), and the parameters of the *communication model* (e.g., the number of concurrent communications allowed).

The partitioning methodology explores the design space (it is based on a genetic algorithm) looking for feasible solutions, supporting also the selection of a heterogeneous multi-processor architecture (what components must be included and how these should be connected) taking into account several issues (degree of affinity, communication cost, processing elements load, concurrency, area, physical costs, etc.). It decides the binding between parts of the behavior and the selected components. Architecture selection and partitioning are influenced by performance requirements, implementation cost, and application-specific issues.

The timing co-simulation methodology considers the proposed heterogeneous multi-processor architecture and a high-level model for the communication media in order to model the system behavior through the behavior of the hardware and software parts. It evaluates the performance of the system by verifying its timing correctness. Moreover, it allows (and suggests) modification in the communication model parameters, and the designer intervention to set different scheduling directives.

The partitioning methodology and the related tool are discussed in Chapter 7, while Chapter 8 analyzes in detail model the methodology and tool related to functional and timing simulations.

Specification transformation

This step involves the specification modifications that can be performed in order to satisfy the design constraints. In detail, it is possible to go back in the flow through this step from several points (dotted lines in Figure 3.7), each of them more costly than the previous one, that is, after the functional co-simulation, after the system design exploration, and after the low-level flow.

The first case arises when the functional co-simulator detects some functional errors in the system behavior, that is, there are some functional errors in the specification itself. In this case, the designer is called to express correctly the desired behavior.

In the second case, the system design exploration step was not able to provide a solution able to satisfy the constraints. To solve this problem, other than trying to change system design exploration parameters (architectural constraints, scheduling directives, and communication model), it is possible to modify the specification by applying some transformations suitable to explicitly show some specification features (e.g., concurrency) that the tools could take into account in the following steps. Such transformations are based on process algebra and their application provides several semantically equivalent OCCAM specifications. A detailed description of such transformations can be found in [99]. If such modifications do not provide any successful response, one or more constraints should be eventually relaxed.

Finally, there is the third case in which, after the low-level step, the obtained implementation does not satisfy some of the constraints. This is the worst situation and the goal of a co-design tool is to avoid it. However, this possibility should be taken into account and so there are different possible actions: to re-execute the low-level flow trying to change its parameters, to re-execute the system design exploration trying to change its parameters, to modify the specification, or to relax some constraints. The focus of this book is on the effectiveness of the high-level flow and so this aspect will not be considered anymore in this work.

3.2.1 Target Architecture

The most obvious way to combine performance, flexibility, and cost efficiency is to take the best from different worlds. By their nature, software implementations on programmable processing elements are preferred to realize maximum flexibility. Tasks, which run inefficiently on general-purpose processing elements, have to be mapped on specific processors or on dedicated co-processors. For example, signal-processing tasks are in general better supported by *DSPs* than by *GPP*, while the opposite is true for control tasks. These considerations lead to the concept of heterogeneous multi-processor system architectures.

However, different application domains (e.g., video, audio, telecom, automotive) have different requirements on speed, power dissipation, computing power, etc. Therefore, an optimal general architecture does not exist. The optimal solution is found by defining a sort of *template architecture* that can be optimized for the specific characteristics of the application domain. Such an architecture is a set of matching hardware and software modules

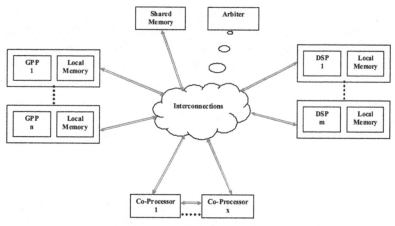

Figure 3.8 Target architecture.

configured according to a specific communication structure. From a suitably defined architecture, different instances can be derived, having the same general structure but differentiated with respect to the number and type of the modules. Thus, each implementation is determined by the application domain and specific application.

The proposed approach improves an existing methodology to cover the case of multi-processor embedded systems. In particular, the proposed extension takes into account different classes of processing elements in terms of both hardware support (*ASIC*, *FPGA*) and programmable processors (GPP, DSP). The target architecture is a heterogeneous one in which, various types of communications link interconnect processing elements and multiple tasks concurrently run on the system. Such a system can be obtained by means of a loosely coupled architecture, using a *low-dimensional regular direct* interconnection topology [45]. The architecture thus identified will reach the maximum flexibility by considering a sort of *MIMD (Multiple Instruction Multiple Data)* paradigm in order to exploit the so-called coarse-grain parallelism, which parcels tasks between different processing elements at an abstraction level where tasks can be matched with the functionalities of the system. This makes it possible to run the same software either on a single-processor system or on a coarse-grain heterogeneous multi-processor architecture and, depending on the constraints, it is possible to explore the design space being aware that the target architecture will be capable of supporting every solution provided by the system design exploration step.

The main differences with respect to the *TOSCA* target architecture (Figure 3.3) are: the presence of multiple heterogeneous processors (each one with its own local memory), a customizable interconnection topology (*bus*, *multiple buses*, *crossbar*, *mesh*, etc. [45]), and an arbitrage mechanism to regulate the accesses to such a network. It is worth nothing that, as in the TOSCA environment, the inter-task communication is hardwired in the case of hardware-hardware interactions. Figure 3.8 shows the architecture template that will be considered as the reference one in the rest of this book.

3.3 Conclusion

In the field of heterogeneous multi-processor embedded systems design, a co-design framework is of critical importance: it enables the designer to optimize the design process and the design itself. This work is an attempt to fulfill the top-level requirements of a co-design environment able to support the concurrent design of embedded systems, eventually subject to real-time constraints implemented exploiting heterogeneous multi-processor architectures.

This chapter has presented the starting point of this project, the *TOSCA* environment (an existing single-processor co-design environment [4, 9, 11, 62]) and the extensions proposed in this book to support, at system-level, co-specification, co-analysis, co-simulation, and system design exploration of heterogeneous multi-processor embedded systems, providing directives to guide low-level co-design tools. Therefore, this chapter has provided a complete overview of the proposed flow, of the related environment (TOHSCA).

4

System-Level Co-Specification

This chapter presents a review of the state-of-the-art techniques used for system-level co-design, focusing on system-level specification languages. The attention is on the design of entire systems implemented on single silicon die, systems-on-chip (*SoC*), but the discussion applies also to different levels of integration (e.g., system on board, or distributed systems).

A system can be modeled at different levels of abstraction: the most used are the architectural and behavioral ones. At the architectural level, the system is represented as an abstract network of interconnected functionalities. The functionalities are typically modeled as black boxes whose interface only is known. This representation captures in a compact way the behavior of the system but does not give any detail on the internal implementation. The architectural level description can be converted into a functionally equivalent behavioral description. This process is currently performed manually by the design team, in some cases by using commercial tools. The behavioral view adds an algorithmic description of the functionalities of the system.

Such description represents the system-level co-specification: it is provided by means of proper specification languages describing the system to be designed and, often, what the performance requirements are. Thus, the system-level co-specification is the entry point to the co-design flow.

To be able to determine characteristics as the execution time of the behavior, the co-specification should be executable: this means that it is most conveniently described in a notation close to a programming language. Such *programs* are often structured as a set of concurrent tasks, and the statements of each task are executed sequentially. In some cases, the tasks themselves may contain several parallel subtasks to support very fine-grain parallelism.

This chapter is structured as follows: Section 4.1 gives an overview of the most representative system-level specification languages, while Section 4.2 describes with some detail the *OCCAM* language [12] as the reference language of the proposed co-design flow. Moreover, it takes into account *SystemC*

[14], a promising standard in this field, showing some similarities between the two languages. Finally, since the co-specification should be represented in a way useful for the automated processing by means of proper tools, one or more *internal models* of representation should be adopted. Therefore, Section 4.3 presents the internal models (*statement-level* and *procedure-level*) used in *TO(H)SCA* to represent the OCCAM co-specification. It is worth noting that the procedure-level internal model, defined and adopted in this book, is suitable to represent the main features of several specification languages (in particular SystemC) enabling the system design exploration methodology developed in this book to be adopted in different co-design environments.

4.1 System-Level Specification Languages

The system-level specification of a hardware/software system usually follows one of the two approaches described below: *homogeneous* or *heterogeneous* specification. With a homogeneous specification, the whole system is described using a single language. This approach simplifies specification and has the advantage that no components of the model are pre-bound to a specific partition, but it poses severe limitations on the simulation efficiency and accuracy because a language powerful enough to express in detail the properties of both hardware and software is often complex and incomplete. In fact, no languages are currently available to efficiently model strongly different and heterogeneous modules. Such a language is argument of current research [16–18]. On the other hand, one major limitation of the homogeneous approach lies in the difficulty of modeling, with acceptable accuracy, the hardware portion of the design and the microprocessor running the software.

The specification languages used by researchers in co-design derive from both *Hardware Description Languages* (*HDLs*) and classical programming languages. A HDL model of the microprocessor is hardly acceptable: at the behavioral level, it does not provide cycle-accurate results that are often necessary; at the *RT* (*Register Transfer*), or gate-level, when available, it requires far too long simulation times.

With a heterogeneous specification, the system modules are modeled using different languages. In this case, the behavior, as well as the structure, of each component can be accurately described by a proper language. This leads to an effective simulation of the stand-alone components but requires co-design environments capable of efficiently integrating such languages into a single, coherent, and executable specification, implementing also the communication among different simulation engines. When the heterogeneous strategy is adopted, the hardware part of the system, described in one of the many HDL, is simulated with commercial tools while the microprocessor running the software program is modeled using ad-hoc languages that provide the required efficiency and accuracy. However, this kind of approach could give rise to an implicit partitioning of the specification, i.e. the language used could be implicitly bound to a particular implementation.

Some of the more recent and widely used languages for system-level specification (homogeneous or heterogeneous) are briefly reviewed in the following:

The most popular hardware specification languages are *VHDL* [19] and *Verilog* [126], developed for the description and simulation of digital circuits. These formalisms can be used to describe the system structurally (i.e., architectural level), functionally, and behaviorally. The functional view, often referred to as register-transfer level or RT level, exploits a number of complex operators, such as adders, multipliers, etc., and some constructs typical of software programming languages, such as conditional statements, loops, functions, etc. At this level of abstraction, the designer can neglect the implementation details of operators and constructs, and can thus concentrate on the functionality. Nevertheless, these languages require an explicit notion of time and operations must be assigned to predefined clock cycles. This means that operation scheduling and operators binding must be performed manually, and in advance, by the designer or the design team.

At a higher abstraction level, the functionality of the design can be captured in a purely algorithmic manner, without explicit assignment of operations to specific clock cycles. Such a model is referred to as behavioral and can be conveniently specified by means of HDLs.

One might argue that HDLs are very appropriate languages for co-design, since it provides a direct route to *Application Specific Integrated Circuits* (*ASIC*) implementation by using commercial high-level synthesis systems, which accept this language as input. However, the problem is that they have been designed for circuit simulation, which tends to make the specifications biased toward solutions less appropriate for software implementation.

The development in co-design goes toward increasing levels of abstraction, and it is therefore expected that this will be reflected in the use of more abstract languages than HDLs for describing heterogeneous systems.

Therefore, as opposite to the HDL choice, several researchers have proposed the use of classical programming languages. One of the classical programming languages, thought also for specification of embedded systems, is *Ada* [20]. The 1995 revision of the language introduced many features, which make it more suitable for embedded (real-time) applications, e.g., protected types that were essentially borrowed from the *Orca* language [21], better interfacing to hardware devices, and improved real-time clock handling.

Thanks to the significant improvements of compilation techniques and of commercial as well as public-domain compilers, software functions are nowadays specified using a broad set of high-level programming languages such as *C* [22, 23], *C++* [24–27], *Java* [28, 29], *MATLAB* [30], and *Esterel* [31–33]. However, this approach could present the same drawback as the HDL one: to bias the specification toward a particular implementation (i.e., in this case the software one).

Therefore, a variety of different formalisms have been proposed and used to capture architectural and behavioral descriptions of a system: graphical models [34, 35] (often defined ad-hoc by *CAD* or *EDA* vendors), different flavors of *Petri Nets* [36], and *CSP* (*Communicating Sequential Processes*, [37, 159]). The major limitation of such formalisms is that they either lack generality or serve as a more abstract front-end for other languages (commercial tools exist that translate a high-level model using these formalisms into VHDL).

For example, an interesting language developed for specifying embedded systems is *Statecharts* [34, 35, 38–40]. It lets the designer describe the behavior graphically as a hierarchy of concurrent finite state machines, which communicate synchronously. An interesting attempt to overcome some of the deficiencies of HDLs is the *SpecCharts* language [42, 43]. It combines the structured task graphs of Statecharts with the programming language-style specifications of HDLs, which is used to describe the behavior of leaf nodes. In [41], it has been shown how the SpecCharts specification could be translated into pure VHDL for simulation and synthesis purposes. Another system specification language that works to high-level of abstraction is *SDL* [46]. It was originally developed for use in the telecommunication area, but could be equally suitable for any distributed system. The reason that makes it appropriate for such systems is that SDL tasks communicate asynchronously, by message passing. This property could also be of importance for heterogeneous implementations, since the various components often run at different clock frequencies, which makes asynchronous communication almost a must, except for very tightly coupled systems. Another language, which is conceptually very similar to SDL, is *Erlang* [47]. The main difference is that the internal behavior of a task in Erlang is written in a functional language, whereas in SDL it is expressed using finite state machines.

In recent years, a number of companies and universities are studying new languages for modeling both hardware and software and, possibly, including the specification of design constraints. Owing to the increasing complexity of mixed hardware/software systems, a desirable characteristic of a specification language is its possibility to be formally verifiable. The most promising proposals have been, in the past few years, *SpecC* [48–51] and *HardwareC* [52]. Recent efforts have led to the definition of languages, such as *SystemC* [14, 53], *SLDL*, *Rosetta* [16–18], and others, whose purpose is to capture different views of the design.

In particular, SystemC seems to be a promising language, and it is candidate to be a standard in the field of system design. This is due to several reasons related to some positive aspects of the language itself and to the notable representativeness, in the field of system design, of the major proposes of such a language.

Finally, the diffusion of SystemC, that is *simply* a class library aimed at extending C++ making it useful for system specification, has lead the way for other languages typically used to specify and design object-oriented applications (e.g., *Unified Modeling Language*, *UML* [15]).

The next Paragraph discusses the reference language adopted in the TO(H)SCA environment and compares it with SystemC as system-level specification language.

4.2 Reference Language

This section analyzes in detail the specification language and the internal representation adopted in the *TO(H)SCA* environment: the *OCCAM* language. Finally, a comparison with SystemC is proposed, showing how some of the main features of this work can be ported to SystemC-based design environments.

4.2.1 OCCAM

The entry point of the *TO(H)SCA* environment is a homogeneous specification based on the *OCCAM* language (namely *OCCAM 2*). OCCAM [12] is a programming language developed by *INMOS* in the 1980s to exploit the parallelism of transputers.

This language offers some features that are extremely attracting for mixed hardware/software specifications, features that have been identified in the literature [54, 55] as optimal for system specification languages. In particular,

- *Hierarchy*: in the OCCAM the hierarchy is expressed decomposing the specification in procedures that represent system functionalities or part of them;
- *Concurrency*: OCCAM processes can be grouped using the *PAR* statement, enforcing parallel execution. The resulting compound process terminates when the longest of its sub-processes terminates. The notion of OCCAM process refers to each single or compound statement of the language.
- *Sequentiality*: OCCAM processes can be grouped using the *SEQ* statement to enforce sequential execution, like in an ordinary programming language. The resulting compound process terminates when the last sub-process terminates.
- *Arbitrage*: OCCAM processes can be executed in mutual exclusion, according to two conditions: the presence of data on a channel (see below) and the value of an ordinary Boolean condition. The *ALT* statement implements this behavior. A priority can be assigned to the sub-processes using the *PRI ALT* statement.
- *Communication*: OCCAM channels provide a mechanism for both intra-process communication and synchronization. Channels are unidirectional, point-to-point blocking channels. The semantics of communication is based on the strict rendezvous protocol.
- *Protocols*: OCCAM channels can transport data organized according to a PROTOCOL definition. OCCAM protocols are similar to *C structures* with the difference that the fields of a protocol are transmitted over a channel sequentially rather than in parallel. Protocols are not supported by the current version of TO(H)SCA.
- *Traditional software constructs*: OCCAM, being a programming language, provides all the constructs of a common programming language (*C*, *Pascal*, *Fortran*, etc.) such as arithmetic, conditionals (*IF*), loops (*WHILE*, *FOR*), and procedures (*PROC*).

The OCCAM language offers a number of other, more complex, features that are currently not supported by TO(H)SCA. In particular, the last version of OCCAM, called OCCAM 3 [12], supports dynamic memory allocation and pointers.

These advanced features, and others (e.g., recursion) have been forbidden in TO(H)SCA, where the language is used as a specification, not as a programming one. This is mostly due to the necessity of preserving the possibility of synthesizing in hardware each OCCAM procedure. The supported subset is summarized in Table 4.1.

Table 4.1 OCCAM subset supported by TOSCA

Construct Class	Supported
Container	PROC, PAR, SEQ
Control	ALT, IF, WHILE
Basic Processes	Assignments, SKIP, STOP
Communication	! (output), ? (input)
Variables	Channels, Variables, Constants
Types	BIT, BYTE, INT

Table 4.2 OCCAM extension for constraints support

Parameter	constr
Area	AREA
Timing	DELAY, RATE
Power	POWER

Though powerful, this subset suffers the limitation that it is not possible to assign a name to a compound process smaller than a PROC, and, consequently, it is not possible to set constraints on portions of the code.

To this purpose, some keywords have been added to the OCCAM language. The keyword TAG, followed by a name, is used to assign a name to the process immediately following the TAG declaration itself. Using these names, the constructs *MAX constr* and *MIN constr* can set constraints (upper and lower bounds) on single processes as well as on groups of processes. The string *constr* specifies the type of constraint (Table 4.2).

OCCAM vs. SystemC

SystemC [14] is a modeling platform (i.e., a C++ class library) that allows heterogeneous systems to be specified using C++. In a SystemC description, the fundamental building block is a process. A process is like a C or C++ function that implements a defined behavior. A complete system description consists of multiple concurrent processes. Processes communicate with one another through signals, and explicit clocks can be used to order events and synchronize processes. Using the SystemC library, a system can be specified at various levels of abstraction. At the highest level, only the functionality of the system may be modeled. Moreover, different parts of the system can be modeled at different levels of abstraction and these models can co-exist during the system simulation.

OCCAM and SystemC have several common aspects. They are both executable imperative specification languages based on the same concepts of concurrency, hierarchy, and communication.

SystemC is, in general, a specification language more expressive than OCCAM; however, when it is used to describe a system at high levels of abstraction (*Untimed Functional Level* or *Timed Functional Level*), many aspects are not necessary and then the language expressiveness becomes similar to OCCAM. Obviously, there will always be a great difference: OCCAM is a procedural language whereas SystemC is an object-oriented one.

In the OCCAM language, concurrency applies explicitly at task or statement level; the hierarchy is expressed decomposing the specification in procedures that represent system functionalities or part of them; communication and coordination are achieved by means of channels, that are ideal point-to-point unidirectional blocking links between procedures. The communication protocol is based on the message-passing model.

In SystemC, concurrency applies implicitly at task level; the hierarchy is based on the concept of modules: a module is a process container where processes are used to describe the module functionality; communication and synchronization at system-level is based on links between modules. The communication over these links is based on the *Remote Procedure Call* protocol.

Going beyond the different names of the language constructs, it is important to note that the basic building blocks are nearly the same. There are groups of basic statements (OCCAM or SystemC statements) that represent system functionalities (or part of them) that can be executed sequentially or in a concurrent manner. Moreover, these groups can communicate by means of logic channels that allow the exchange of data and the synchronization making the groups able to cooperate.

These similarities have been exploited defining a procedure-level internal model, shown in the next paragraph, that is suitable to represent specification expressed both in OCCAM and in SystemC. This important feature enables the system design exploration methodology developed in this book to be adopted in different co-design environments.

4.3 Internal Models

Starting from the *OCCAM* specification, the *TOHSCA* environment makes use of two different internal models to represent in a *processing-oriented* way the relevant aspects of the specification. The models present two different granularities: a statement-level model, strictly related to the OCCAM language (used also in TOSCA), and a procedure-level model, suitable for representing specifications derived from different specification languages.

When a new project is loaded into the TO(H)SCA environment, an *OCCAM Parser* processes the OCCAM source file and builds the statement-level internal representation based on the *aLICE* library, a C++ classes library developed with the purpose of providing the basic blocks to build OCCAM internal models and to integrate new applications in the TO(H)SCA environment.

After co-analysis, co-estimations, functional co-simulation, and load-evaluation steps, the procedure-level internal model, annotated with all the useful information on the system gathered in the previous steps (Figure 3.7), is provided by the co-simulator to the partitioning tool. The procedure-level model allows dominating the system complexity during the partitioning step and it is applicable to several specification languages, making the partitioning tool suitable for different environments. Moreover, to enhance generality, the exchange format of this model is based on the VCG format [84], a third-party format that can be managed and visualized with third-party open source tools.

In the case of TO(H)SCA, the VCG file generated by the co-simulator enables the partitioning tool to build the procedure-level internal model. In the following, the statement-internal model is analyzed with some detail, while the innovative procedural-level model is formally defined.

4.3.1 Statement-Level Internal Model

The structure of the *TO(H)SCA* statement-level internal model is organized as a list of trees plus a number of lists for global references and additional information (Figure 4.1). Each tree is the model of a single OCCAM procedure.

A tree representing a procedure is built according to the following basic rules and schemes. A procedure node (PROC) has a single child (its body) and a list of the formal parameters (Figure 4.2).

A container node (SEQ, PAR, IF, WHILE, ALT) has a list of the local variables and a list of one or more child. For the SEQ and PAR containers, a child is a *process*. For the IF container, a child is the pair *condition-process* (the WHILE process is analogous to an IF with only a child). Finally, for the ALT container, a child is the triplet *condition-input-process* (Figure 4.3).

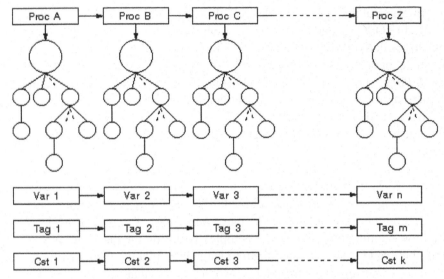

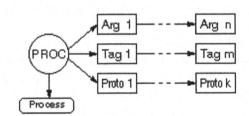

Figure 4.1   Structure of the OCCAM internal model.

Figure 4.2   Structure of PROC internal model.

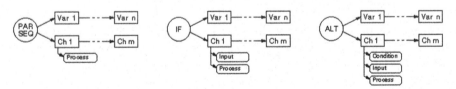

Figure 4.3   Structure of PAR, SEQ, IF, and ALT internal models.

The expressions are represented as a *syntax tree* whose leaves can be either literals or variable references. The literals appear directly in the expressions while variables are referenced through a special object. The figure below shows a typical syntax tree (Figure 4.4).

Further details on the *OCCAM Parser*, the *aLICE* library, and the internal model discussed above can be found in [10].

a + (b / c[n+1])

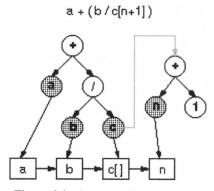

Figure 4.4   Structure of a syntax tree.

4.3.2 Procedure-Level Internal Model

The *TOHSCA* procedure-level internal model is able to capture information related to the computational elements present in an imperative, eventually *object-oriented*, specification and the relationship between these. Such a model, called *Procedural Interaction Graph (PING)*, based on the *Procedural Call Graph (PCG,* [85]), is a formalism composed of nodes (i.e., instances of methods or procedures) and annotated edges (i.e., method or procedure calls, communication, or synchronization operations) that provide information on the relationships between the nodes and on the data exchanged (e.g., size, profiling, etc.). The *PING* is optimal to represent a coarse grain view of the system that takes into account communications, synchronizations, and concurrency issues. In the following, some definitions provide a formalization of the model that is applicable to imperative object-oriented (*OO*) specification languages as well to not object-oriented (*NOO*) ones.

Basic definitions

Definition 1
LN = {VI, VO, VT} is the adopted specification language where:

- *VI* is the set of valid identifiers of *LN*;
- *VO* is the set of valid operators of *LN*;
- *VT* is the set of valid types of *LN*.

Definition 2
VS is the set of valid specifications that can be expressed with *LN*.

Definitions related to *PING* nodes

Definition 3

The triple $m_{i,j} = \langle q, k, c \rangle$ is the *method j* of class c_i (Definition 5) where:

- $i, j \in N$;
- $q \in VI$ is the method name;
- $k \in MQ = \{B, NB\}$ is a method qualifier where B indicates a *blocking* method (i.e., the caller waits for the callee completion) while NB indicates a *non-blocking* method (i.e., the execution of the caller proceeds concurrently with the callee);
- $c \in VS$ is the method body.

The term *method* indicates both a method in the OO sense and a procedure in the NOO one.

For NOO languages, all the methods belong to a fictitious class 0.

Definition 4

The pair $d_{i,j} = \langle q, z \rangle$ is the *member data j* of the *class* c_i (Definition 5) where:

- $i, j \in N$;
- $q \in VI$ is the data name;
- $z \in VT$ is the data type.

Definition 5

The triplet $c_i = \langle q, MT_i, DT_i \rangle$ is the *class i* where:

- $i \in N$
- $q \in VI$ is the class name,
- MT_i is the set of the class methods,
- DT_i is the set of the class member data.

For NOO languages, the only class with methods is a fictitious one (*class 0*). Therefore, MT_0 is the set of all the methods, and DT_0 is the set of global variables. Other classes c_i can exist if and only if $MT_i = \{\emptyset\}$ (i.e., they are typically called *structures*).

Definition 6

CL is the set of the classes c_i declared in the specification: $CL = \bigcup_i c_i$

Definition 7

MT is the set of the methods of all the classes: $MT = \bigcup_i MT_i$

For NOO languages, $MT = MT_0$ is the set of all the procedures.

Definition 8

$o_i^k \in VI$ (where $i, k \in N$) is the instance k of a class c_i.

Definition 9

$OB = \bigcup_{i,k} o_k^i$ is the set of all the instances of all the classes.

Definition 10

$f_{k,j}^i$ (where $i, j, k \in N$) is the instance of the method $m_{i,j}$ of the instance k of class c_i.

For NOO languages that allow multiple instances of a method (i.e., procedure), $f_{k,j}^i$ is the instance k of the method $m_{0,j}$.

Definition 11

$MI = \bigcup_{i,k,j} f_{k,j}^i$ is the set of all the method instances in the specification.

Definitions related *to PING edges*

Definition 12

The tuple $t_i = \langle s, w, d, g, f \rangle$ is the data transfer i where:

- $s, d \in MI$, with $s \neq d$, are respectively the source and the destination method instances of the data transfer;
- $w \in TQ = \{MC, CS\}$ is a data transfer qualifier where PC indicates a method call (i.e., a procedure call) while CS indicates a communication (or synchronization) operation;
- $g \in N$ is the exchanged data size in bit; and
- $f \in Z$ is the average number of times that the data exchange occurs (that can be evaluated by means of profiling).

Definition 13

$DT = \bigcup_i t_i$ is the set of all the data transfers present in the specification.

Definitions related to *PING*

Definition 14

The pair $G = \langle MI, DT \rangle$ is the graph of a given specification, where the graph nodes are the instances of method, and the graph edges are the data transfers.

A graphical representation of such a graph can be obtained, for example (Figure 4.5), by associating a node for each instance of a method (the node

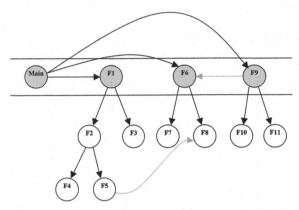

Figure 4.5   Internal model graphical representation.

is gray if the method is non-blocking) and an arrow for each data transfer (the arrow is gray if the data transfer is a communication/synchronization operation). The non-blocking methods are represented at the same level in order to provide an intuitive graphical representation of the possible concurrency in the system (this could be however limited by communication and synchronization issues).

The example of Figure 4.5 shows a graphical representation of a procedure-level internal model, where the instance of the *Main* method calls the instances of non-blocking methods *F1*, *F6*, and *F9*. The other instances of the method make only blocking calls, while instances *F5–F8* and *F6–F9* make *CS* (Definition 12) data transfers (gray arrow).

In the case of the *OCCAM* language, the nodes are OCCAM procedures (they can be explicitly called to be executed concurrently) and the arcs are procedure calls or channel communication/synchronization operations. In the case of *SystemC*, the nodes are instances of a method (classical method, *SC_METHOD*, or *SC_THREAD*) and the arcs are method calls or communication/synchronization operations by means of *RPC* or *sensitivity lists*.

4.4 Conclusion

This chapter has presented a review of the state-of-the-art techniques used for system-level co-design, focusing on system-level specification languages. Successively, the TO(H)SCA reference language and the internal models (*statement-level* and *procedure-level* internal model) used to represent the specification have been introduced, in particular, it has been shown how the procedural-level internal model, defined and adopted in this book, is suitable to represent the main features of several specification languages enabling the proposed system design exploration methodology to be adopted in different co-design environments.

The next chapters describe in detail the different tools that implement the methodologies developed to support the proposed system-level co-design flow. Such tools provide information that is annotated on the presented models allowing the partitioning tool (Chapter 7) to work on a compact view of the system functionalities, while providing all the useful information.

5

Metrics for Co-Analysis

Co-analysis is a task that, starting from system-level specification, provides quantitative information useful to make system-level decisions such as architectural selection and partitioning. The underlying idea is that the performance metrics of a final design can be related to the properties of the specification itself. Therefore, the core of this task involves the identification and evaluation of functional and structural properties of the specification, which could affect the design performance on different architectural platforms.

The idea of using algorithms properties to provide design guidance for general-purpose architecture is not new. For example, the *Amdahl's law* [2] is an attempt to indicate the amount of speedup attainable on an algorithm by using a parallel architecture; locality issues and the 90/10 *rule of thumb* [92] have been fundamental to the introduction of cache memories in modern computer architectures. In the *VLSI DSP* domain, the qualitative observation that many signal processing algorithms are regular has motivated the development of *systolic* and *wavefront* arrays [93]. The analysis of the algorithms properties is actually very important also in the field of parallel computing research.

Therefore, this book adopts such an approach as the first step of the proposed co-design flow (Figure 5.1). Co-analysis aims at obtaining as much information as possible about the system by statically analyzing the specification. More in detail, the goal of this step is to statically detect the best processing element for the execution of each system functionality. In fact, in the co-design of heterogeneous multi-processor embedded systems, the choice of the processing elements that constitute the final architecture should be based not only on performance issues but also on the features of the tasks that they should execute.

The proposed analysis provides a set of data expressing the *affinity* of a functionality toward a type of processing element (*GPP*, *DSP*, and *ASIC/FPGA*). To characterize the specification, two subtasks should be

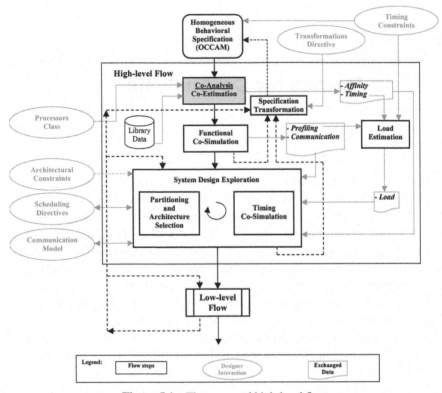

Figure 5.1   The proposed high-level flow.

performed: an architectural analysis of the existing processing elements, in order to detect their relevant features, and the definition of a set of patterns able to identify subsets of the specification that could exploit the identified architectural features. The gathered data support the definition of a set of metrics providing meaningful indications useful to make design choices. It is worth noting that such tasks should be performed once for all and modified only when technology innovations give rise to new relevant kinds of processing elements.

This chapter is structured as follows: Section 4.1 presents a characterization of the main processing elements architectural features, while Section 4.2 presents the proposed model and the methodology developed based on such a characterization. Moreover, such section describes the validation process showing the experimental results obtained by applying the presented methodology to a meaningful test suite. Finally, Section 4.3 draws some conclusions on the overall results obtained in this chapter.

5.1 Characterization

The first relevant subtask to be performed, in order to enable the application of the proposed methodology, is a proper characterization of the main executors adopted in the field of embedded design. An architectural analysis of such processing elements is necessary to detect their relevant features to be exploited for the execution of the most appropriate functionalities.

A first characterization discriminates between *processor-like* and *ASIC-like* executors. The former is equipped with a more complex control unit so it is quite independent of the retrieval and management of data. The latter is more suitable in a *co-processing architecture*, i.e., where it acts as a co-processor for a main device, performing only particular tasks.

Processor-like devices present a fixed-width data-path and can perform easily several different sequences of operations: they are optimal for low-regularity applications, i.e., applications in which a variation in the data implies a notably variation in the kind of operations to be performed, and the control flow is difficult to be predicted.

ASIC-like devices are instead suitable for high-regularity applications in which the operation to be executed are repetitive and nearly independent of data. Moreover, such operations should be executable concurrently in order to exploit to the best the physical parallelism provided by such kind of devices.

Therefore, processor-like devices better exploit the available resources by means of programmability for low-regularity tasks, while ASIC-like devices are more promising for regular tasks especially if there is a large amount of exploitable parallelism.

Such preliminary considerations represent a first guideline toward an effective association between functionality and category of executors. Such categories are listed in the following together with a brief description:

- **General Purpose Processor (*GPP*)**: processors without any specific characterization.
- **Digital Signal Processor (*DSP*)**: processors tailored to digital signal processing applications and so they present a loss of generality with respect to GPP and a higher cost, but they provide a better performance in the execution of a particular set of instructions [90]. Moreover, they are typically equipped with several high-bandwidth I/O ports and certain amounts of on-chip memory.
- **Application Specific Integrated Circuit (*ASIC*)**: integrated circuits (i.e., generally more performing) for specific applications. Their design and development costs are very high so they are affordable only for large production volumes.

- **Field Programmable Devices** (*FPD*): arrays of logic blocks with programmable interconnections that define the performed functionality. Such programming can be executed one or more times depending on the particular technology. They represent a tradeoff between processors and ASICs with respect to performance, flexibility, and cost. In particular, this book takes into consideration the *Field Programmable Gate Array* (*FPGA*, [86–88]).

The remaining of this section presents a deeper analysis in order to detect exploitable architectural features about the executors considered above.

5.1.1 GPP Architectural Features

GPPs are designed to be useful in several contexts and so it is difficult to detect particular architectural features that strongly identify a *GPP-suitable* application. They are typically adopted as control elements and I/O manager, but they are also useful (especially if provided of additional components like a floating-point unit) for general computations. Typical data managed by a GPP are integer with standard size (i.e., 32/64 bit) and single and double precision floating point numbers. All the technology enhancements (e.g., pipeline units, cache memories, etc.) applied to such devices aim at improving the average performance. However, such enhancements could generate predictability issues when the devices are used in the field of real-time applications. For complex systems, that use an operating system, this is typically executed by a GPP that acts as a manager for processes, memory, and I/O.

Therefore, the functionalities more suitable for a GPP executor are those that provide management of complex communication protocols and, in general, all those that present a complex and irregular control flow of execution.

5.1.2 DSP Architectural Features

DSPs are high-performance executor for digital signal applications. This implies a loss of generality with respect to GPP and, in general, a higher cost; conversely, they provide a better performance in the execution of a particular set of instructions [90]. For example, typical DSP operations are represented by regular (i.e., repetitive) computations on fixed length arrays (e.g., filtering, correlation evaluation, etc.).

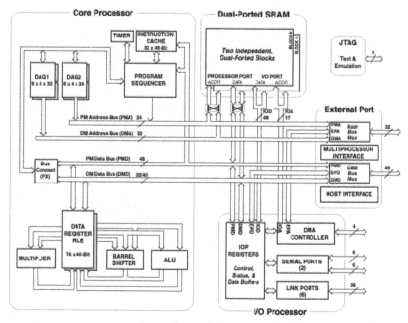

Figure 5.2 SHARC Internal Architecture.

Architectural features included in a DSP are those that allow concurrent load of multiple operands, concurrent execution of sums and multiplications, and fast access to sequential memory space (i.e., *array*) while keeping, however, particular attention to predictability issues.

DSPs internal architectures have been designed carefully to support such features. As an example, the *Super Harvard Architecture* (Figure 5.2) adopted in the *Analog Devices SHARC DSP* [91] is considered.

Such an architecture is based on a dual internal bus that allows the concurrent fetching of instructions and operands and avoids, as while as possible, memory access conflicts. Moreover, keeping in an internal memory the instructions related to the loops, it is able to fetch two operands at a time. Let us consider the simple code in Figure 5.3 to show other specific characteristic of the SHARC (and in general to all the DSPs).

This code is composed of a loop with its body. In the body, there are a sum and a multiplication that involve two arrays. The first enhanced architectural feature is related to the management of the loop. The SHARC, after a setup phase, is able to manage the loop counter (increment and test) automatically and concurrently with the body execution (*Zero Overhead Looping*).

```
...
result = 0;
for ( n = 0 ; n < LENGTH ; n++ )
        result += x[n] · y[n];
...
```

Figure 5.3   MAC example.

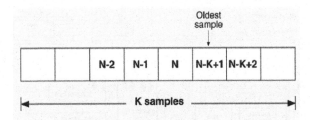

Figure 5.4   A circular buffer of length *k*.

Two *DAG* units (*Data Address Generator*), one for each bus (see Figure 5.2), are dedicated to automatically update the address of the index register: the index value of the array is evaluated concurrently with the fetch of the current data. If the updated index value exceeds a limit, the DAG adjusts the index so that it wraps. This occurs in the same clock cycle as the read, or write, operations. Moreover, such DAG provides two important features: the *Circular Buffering* and the *Bit Reverse Addressing*.

The former is useful, for example, in applications that operate on vectors of samples. The next input vector is derived from the last input vector by shifting all elements of the last vector. The oldest entry in the vector is shifted out and a new entry is shifted in. Rather than actually moving the data, circular buffers are employed. By updating an index *modulo* the buffer length, the oldest entry can be conveniently replaced by the newest one (Figure 5.4).

The latter is able to enhance the performance of very common and critical functionalities, like the *Fast Fourier Transform* (*FFT*). One requirement during FFT computation is the *bit-reversed addressing* (the FFT code related to this feature is shown in Figure 5.5). In bit-reversed addressing, the bit pattern of a vector index value is reversed. The SHARC can automatically perform this bit reversal while accessing data.

```
...
n = nn << 1;
j = 1;
for ( i = 1; i < n; i += 2 )
{
        if ( j > i )
        {
                swap( data[ j ] , data[ i ] );
                swap( data[ j + 1 ], data[ i + 1 ] );
        }
        m = n >> 1
        while ( m >= 2 && j > m )
        {
                j -= m;
                m >> 1;
        }
        j += m;
}
...
```

Figure 5.5   Bit reversing in the FFT.

Other than the features considered by now, it should be noted that the ALU of the SHARC works with integer, float, or fixed length data types with a constant clock cycles number: this is very important for predictability issues.

Finally, the evaluation of expressions like the one presented in Figure 5.3 (result += $x[n]$ * $y[n]$) is executed by means of a single instruction, called *MAC (Multiply Accumulate)*, that allows a DSP to evaluate multiplications concurrently with sums, subtractions, or average evaluations.

Detailed examples, which quantify the gain of performance obtained by a DSP with respect to a GPP when they are involved in the execution of code that presents one or more of the features presented above, are shown in [90].

5.1.3 ASIC-like Devices Architectural Features

This paragraph considers the main common features of *ASICs* and *FPDs* as dedicated executors. The former are ad-hoc IC for specific functionalities: their architectural features are strictly related to the specification and, during the synthesis, a trade-off between area (i.e., cost) and performance should be considered. Their design and development costs are very high so their need should be evaluated carefully.

The FPD are arrays of logic blocks with programmable interconnections to implement the performed functionality. Such programming can be executed one or more times depending on the particular technology [89]. FPDs are cheaper and more flexible than ASICs, but they are generally less performing.

The goal of this paragraph is to identify a set of features that allow an early identification of functionalities able to exploit *ASIC-like* devices. The most relevant features are listed below.

- **Non-standard data-path**: a mismatch between application data-path requirements and those presented by the processor data-path could lead to inefficient use of processor resources. For example, most media processing involves computing on nibble or byte level input data and the wide data-path (32/64 bit) of general-purpose processors is an inefficient match for such computations.
- **Bit manipulation**: because of the previous point, ASIC-like devices are more suitable to perform bit manipulation operations (shifting, Boolean operators, etc.). Moreover, bit-oriented functionalities would underexploit any complex processing unit, other than the width data-path.
- **Pseudo static data:** if some of the computation inputs are either static or change infrequently, then such factors can be considered to specialize the data-paths to either decrease operation latency or improve power/area use. An example of this is the use of constant co-efficient multipliers in most signal processing filtering applications.
- **Regularity and concurrency:** repeated operations of similar types on large regular data sets are an ideal candidate for ASIC-like implementations. Regularity in operations imposes less demand on the control unit complexity better exploiting the available resources (i.e., to use area for computation, not for computation management!). Moreover, the fine-grained parallelism typically provided by such devices is a good match for these kinds of computations.

Therefore, computation intensive functionalities (with high data throughput requirements) that exhibit some of the characteristics listed above are optimal candidate to be implemented on ASIC-like devices.

5.2 The Proposed Approach

This section shows the co-analysis methodology adopted in TOHSCA with the goal of obtaining as much information as possible about the system, analyzing the specification in a static manner. Considering the architectural features identified in the previous section, the next paragraph provides the definition of a set of patterns able to identify subsets of the specification that match some executor features, and a set of metrics that quantify such matching.

Finally, these metrics are properly combined in order to build a global metric (the *affinity*) able to suggest the best processing elements for the execution of each system functionality.

With respect to previous attempts to perform similar analysis, the presented one is more general and accurate. For example, in [89] the efficiency of GPPs and FPGAs is evaluated only with respect to the exploitation of the available area. For such purpose, it considers the following features.

- **Granularity**: it is the difference between the native executor data-path and the average size of the data.
- **Context**: it is the difference between the cardinality of the *executor instruction set* and the number of instructions used by the application.

With respect to the variation of such features, it evaluates the *spatial efficiency* of a device.

In [94], the authors create a methodology that fully characterizes any algorithm with respect to the elements of its structure that affect its implementation. Such methodologies are based on the definition of seventeen properties that are gathered into groups such as *size, timing freedom, uniformity, concurrency, temporality, spatial locality, regularity, cyclic properties*, and *control flow properties*. The identified groups are meaningful, however only a few of them are supported by an effective and operative quantification approach, and when such a support is provided, the metrics defined are strictly bounded to high-level synthesis issues (as an example, the methodology is used to estimate the implementation area of a custom ASIC).

A co-design oriented work is instead the one presented in [95] where the concept of *hardware/software repelling* is used to drive an hw/sw partitioning algorithm. The approach is based on the analysis of the system functionalities, detecting a set of features that suggest a repelling of certain functionalities toward certain type of implementation. Unfortunately, the work considers only one kind of software executors, and the set of features considered is not clearly defined. However, this work is a pioneer that should be taken into consideration.

Finally, [96] represents the work more similar to the one presented in this chapter. In fact, it considers multi-processor systems synthesis starting from an object-oriented specification, and it analyzes subsets of such a specification in order to detect features that allow marking them as *control dominated*, *data transformation dominated*, or *memory access dominated*. However, it does not consider dedicated hardware devices (it considers only GPPs, *microcontrollers*, and DSPs), works with a too coarse granularity level (whole classes and not single methods), and poorly defines the metrics to be used within the methodology.

The following paragraphs show in detail the proposed methodology and the model on which it is based. They provide formal definitions of the introduced metrics and the global metric, describing the developed tool and the validation process of the proposed approach.

5.2.1 Model and Methodology

This paragraph shows in detail the proposed methodology and the model on which it is based. They provide formal definitions of the proposed global metric, i.e., the *affinity*, based on the definition of a set of metrics, i.e., a set of indicators that quantify the structural and functional features of each system functionality.

Problem definition

The previous section has shown a taxonomy of the main executors and has identified several features that allow the identification of a matching between functionality and classes of processing elements. However, in order to be useful, such a taxonomy should be supported by the possibility of identify portions of the specification that present a determined feature and to quantify the matching between each functionality and each executor class, where by functionality we intend each single *instance of method* (Definition 10, Section 4.3.2).

The choice of such a granularity level, at which the proposed methodology is intended to work, is related to the strategy adopted by the partitioning tool, as detailed in Chapter 7. The final goal of the presented approach is to provide a global metric, called *affinity* defined, as follows:

Definition 15

The affinity $A_m = [A_{\mathrm{GPP}_m} \; A_{\mathrm{DSP}_m} \; A_{\mathrm{HW}_m}]$ of $m \in \mathrm{MI}$ is a triplet of values in the interval [0, 1] that provides a quantification of the matching between the structural and functional features of a functionality and the architectural features for each one of the considered executor classes (i.e., GPP, DSP, and ASIC/FPGA). An affinity of 1 toward an executor class indicates a perfect matching, while a 0 affinity indicates no matching at all.

In the following, the functional and structural features considered in the affinity are formally defined and the co-analysis model and methodology are described in detail.

Model

The specification should be statically analyzed to collect information allowing the classification of each functionality with respect to the features that characterize each possible executor. The classification is based primarily on the data involved in the execution of a functionality and on its structural properties. Moreover, several properties oriented to particular classes of executors are considered. In the following, a set of metrics is properly defined providing a model for the classification of the specification.

Data-oriented metrics

The goal of these metrics is to take into account the type of data involved in the execution of a given functionality. They are formally defined in the following.

Definition 16
BT is the set of built-in types of the adopted specification language (e.g., integer, real, bit, character, string, etc.).

Definition 17
VR is the set of the *one-dimensional array types* derived from the types of BT.

Definition 18
MX is the set of the *bi-dimensional array types* derived from the types of BT.
    Considering also *user-defined* heterogeneous types and remembering Definition 1 and Definition 6 (Section 4.3.2), it is possible to write: $VT = BT \cup VR \cup MX \cup CL$.

Definition 19
For each $m \in MI$, D^m is the set of the variable declarations made in m.

Definition 20
D_z^m is the set of declarations $d \in D^m$ related to the same type $z \in VT$.

Definition 21

By varying $z \in$ VT, D_z^m allows, for each $m \in$ MI, the evaluation of a set of metrics called *Data Ratio*: they are defined as the ratio between the number of declarations of a determined type with respect to the total number of declarations made in m:

$$\mathrm{DR}_z^m = \frac{|D_z^m|}{|D^m|}$$

Structural metrics

The goal of these metrics is to identify the structural properties of a functionality focusing on the analysis of the control flow complexity. They are formally defined as follows:

Definition 22

S_m is the number of source lines of $m \in$ MI.

S_{Lm} is the number of source lines of $m \in$ MI that contains loop structures (e.g., while, for, etc.).

S_{Cm} is the number of source lines of $m \in$ MI that contains conditional branches (e.g., if, case, etc.).

Definition 23

The *control flow complexity* of $m \in$ MI is

$$\mathrm{CFC}_m = \frac{S_{Lm} + S_{Cm}}{S_m}$$

The value of such a metric is increased by variations in the execution flow due to decision points (i.e., loops and branches); therefore, a linear sequence of instructions has zero control flow complexity.

Definition 24

The *loop ratio* of $m \in$ MI is

$$L_{Rm} = \frac{S_{Lm}}{S_m}$$

Such a metric discriminates between computational and control oriented functionalities. Moreover, high LR values indicate the possibility to exploit a spatially limited computational unit by means of a compact implementation and a strong component reuse.

DSP-oriented metrics

The goal is to identify functionalities suitable to be executed by a DSP by considering those issues that exploit the most relevant architectural features of such executor class: circular buffering, MAC operations, and Harvard architecture.

Circular buffering

The goal is to identify subsets of the specification that access a linear data structure (one-dimensional array, row or column of bi-dimensional array). The use of a circular buffer is identified, more or less explicitly, by portions of code that try to shift an array of one or more position.

The examples in Figure 5.6 show three situations that could arise, presenting different degrees of interest characterized by the presence of a container loop and by the levels of re-indirection: strong and very explicit (a), weak and less explicit (b) (c) circular buffering exploitation.

The following definitions formalize the proposed approach.

Definition 25

S_{SC_m} is the number of source lines of $m \in \mathrm{MI}$ that contains expression of the form:

$$v[i] = v[I \pm K]$$

where:

- $v \in \mathrm{VR}$ is a vector (or a row/column of a matrix);
- $i,\ K \in Z$, and K is a constant value.

...	...	...

for (n=0; n<10 ; n++)
    v[n] = v[n-1];
 	
...
	v[3] = v[2];
	v[2] = v[1];
	
	...
		for (n=0; n<10 ; n++)
		{
		    temp = v[n];
		    ...
		    v[n-1]= temp;
		    ...
		}
		...

(a)	(b)	(c)

Figure 5.6 Examples of circular buffering.

The *degree of strong circularity* of $m \in$ MI is:

$$S_{\text{CD}m} = \frac{S_{\text{SC}m}}{S_m}$$

Definition 26
$S_{\text{WC}m}$ is the number of source lines of $m \in$ MI that contain expressions with:

$$v[K] = f(v[i]) \quad \text{or}$$

$$q = f(v[i])$$

where:

- $v \in$ VR is a vector (or a row/column of a matrix);
- $i, K \in Z$ and K is a constant value;
- $q \in VI$ is an identifier,
- $f(v[i])$ is a generic expression that involves $v[i]$.

The *degree of weak circularity* of $m \in$ MI is:

$$W_{\text{CD}m} = \frac{S_{\text{WC}m}}{S_m}$$

Multiply accumulate

The goal is to identify subsets of the specification that express a particular mix of operations (i.e., a sum and a multiplication) that a DSP can perform concurrently.

Definition 27
$S_{\text{SM}m}$ is the number of source lines of $m \in$ MI that contains, inside a loop, an expression with the following structure:

$$s_1 = s_1 + s_x s_y$$

where s_1, s_x, and $s_y \in VI$.

The *degree of strong MAC* of $m \in$ MI is:

$$S_{\text{MD}m} = \frac{S_{\text{SM}m}}{S_m}$$

Definition 28
$S_{\text{WM}m}$ is the number of source lines of $m \in$ MI that contains, outside of a loop, an expression with the following structure:

$$s_1 = s_1 + s_x s_y$$

where s_1, s_x, and $s_y \in \text{VI}$.

The *degree of weak MAC* of $m \in \text{MI}$ is:

$$W_{\text{MD}m} = \frac{S_{\text{WM}m}}{S_m}$$

Concurrent memory accesses

The goal is to identify subsets of the specification able to exploit concurrent memory accesses to instructions and data, as provided by *Super Harvard* architecture [90].

Definition 29

S_{SH_m} is the number of source lines of $m \in \text{MI}$ that contains, inside a loop, an expression with the following structure:

$$v[i] \; op \; w[i] \quad \text{or}$$

$$q \; op \; w[i]$$

where:

- $v, w \in \text{VR}$
- $op \in \text{VO}$ and it is different from =.

The *degree of strong Harvard* of $m \in \text{MI}$ is:

$$\text{SHD}_m = \frac{S_{\text{SH}m}}{S_m}$$

Definition 30

$S_{\text{WH}m}$ is the number of source lines of $m \in \text{MI}$ that contains, outside of a loop, an expression with the following structure:

$$v[i] op \; w[i] \quad \text{or}$$

$$q \; op \; w[i]$$

where:

- $v, w \in \text{VR}$
- $op \in \text{VO}$ and it is different from =.

The *degree of weak Harvard* of $m \in$ MI is:

$$W_{\mathrm{HD}m} = \frac{S_{\mathrm{WH}m}}{S_m}$$

GPP oriented metrics

The goal is to identify functionalities that significantly rely on operations that involve conditional dependent control flows, complex data structures, and complex I/O management.

Definition 31
The *Conditional Ratio* of $m \in$ MI is:

$$\mathrm{CR}_m = \mathrm{CFC}_m - \mathrm{LR}_m$$

where:

- CFC_m is the *Control Flow Complexity* (Definition X);
- LR_m is the *Loop Ratio* (Definition X).

Definition 32
$S_{\mathrm{I/O}_m}$ of $m \in$ MI is the number of source lines of m that contains I/O operations (e.g., read, write, etc.).
   The I/O *Ratio* for $m \in$ MI is:

$$\mathrm{I/OR}_m = \frac{S_{\mathrm{I/O}m}}{S_m}$$

Definition 33
The *Structure Ratio* for $m \in$ MI is:

$$\mathrm{STR}_m = \frac{\sum\limits_{z \in CL} |D_z^m|}{D^m}$$

ASIC-like Oriented Metrics

The goal is to identify regular functionalities that significantly rely on operations that involve bit manipulation. Therefore, in addition to some of the previous defined concepts (i.e., LR, and DR_{bit}^m), the following is defined.

Definition 34
S_{BM_m} of $m \in$ MI is the number of source lines of m that contain bit manipulation operations (e.g., *and*, *or*, *xor*, etc.).

The *bit manipulation Ratio* for $m \in \text{MI}$ is:

$$\text{BMR}_m = \frac{S_{\text{BM}_m}}{S_m}$$

Methodology

The information gathered by means of the metrics previously defined is organized in a global metric that allows a straightforward characterization of a functionality with respect to each possible executor. Such a global metric, called *affinity* (Definition 15), is operatively defined in the following.

The Affinity

The affinity of a functionality (Definition 15) can be expressed as a normalization function applied to linear combinations of the metrics, with weights that depend on the considered executor class. To determine the proper function and weights, let us define the main dependencies between the metrics and the characteristics of each executor class.

Intuitively, the affinity toward a *GPP* executor depends primarily on:

- the I/O *Ratio*;
- the *Conditional Ratio*;
- the *Structure Ratio*;
- the number of declared variables of *GPP* compatible type (e.g., integer, float, character, string, etc.).

The affinity toward a *DSP* executor primarily depends on:

- the *degrees* (both strong and weak) of *circularity*, *Harvard*, and *MAC*;
- the *Loop Ratio*;
- the number of declared variables of *DSP* compatible built-in type (e.g., integer, fixed point, float, etc.).

The affinity toward a HW executor depends on:

- the *Loop Ratio*;
- the *BIT Manipulation Ratio*;
- the number of variables of bit type.

Therefore, it is possible to evaluate the affinity for $m \in \text{MI}$ as follows:

$$A_m^T = f\left(W \cdot C_m^T\right)$$

where

$$A_m = \begin{bmatrix} A_{\mathrm{GPP}_m} & A_{\mathrm{DSP}_m} & A_{\mathrm{HW}_m} \end{bmatrix}$$

$$C_m = \begin{bmatrix} \mathrm{SCD}_m & \mathrm{WCD}_m & \mathrm{SHD}_m & \mathrm{WHD}_m & \mathrm{SMD}_m & \mathrm{WMD}_m \\ \mathrm{IOR}_m & \mathrm{CR}_m & \mathrm{LR}_m & \mathrm{BMR}_m & \mathrm{DR}_{bit}^m \\ \displaystyle\sum_{z=\mathrm{char,\ string}} \mathrm{DR}_z^m & \displaystyle\sum_{z=\mathrm{int|real}} \mathrm{DR}_z^m & \mathrm{STR}_m \end{bmatrix}$$

$$W = \begin{bmatrix} w_{1,1} & \cdot & \cdot & \cdot & w_{1,14} \\ w_{2,1} & \cdot & \cdot & \cdot & w_{2,14} \\ w_{3,1} & \cdot & \cdot & \cdot & w_{3,14} \end{bmatrix}$$

The weights of the matrix W are set to 1 when the associated metric is meaningful for a given executor class, 0 otherwise. The resulting matrix is the following:

$$W = \begin{bmatrix} 0 & 0 & 0 & 0 & 0 & 0 & 1 & 1 & 0 & 0 & 0 & 1 & 1 & 1 \\ 1 & 1 & 1 & 1 & 1 & 1 & 0 & 0 & 1 & 0 & 0 & 0 & 1 & 0 \\ 0 & 0 & 0 & 0 & 0 & 0 & 0 & 0 & 1 & 1 & 1 & 0 & 0 & 0 \end{bmatrix}$$

The choice to use only binary weights is related to the meaning of the metrics themselves: when a metrics is 1, it means that the considered functionality is completely tailored to the considered executor. For example, if for a given functionality $SMD=1$ (*Strong MAC Degree,* Definition 27), this means that all the lines of the functionality source code contain a MAC operation and so the functionality exploits to the best a DSP.

In this way, the affinity represents the sum of all the contribution determined by each relevant metric. Since such a sum could be greater than one, a function should be applied to obtain an affinity normalized in the [0, 1] interval. Such a normalization is needed to allow a direct comparison between the affinity values related to different executors and to allow the choice of the best executor for each functionality.

The adopted strategy is related to the use of a proper normalization function: one that provides values internal to a bounded interval. We choose the *arctangent* function because it is limited to the interval $[-\pi/2, \pi/2]$ when x varies from $-\infty$ to ∞ so, to normalize the affinity in the interval [0, 1], it should be scaled of a $\pi/2$ factor. Moreover, to take into account that a value of 1 for a single relevant metric means a strong matching between the functionality and the executor, a proper coefficient is multiplied by x in order to obtain an affinity equal to 0.9 in correspondence with $x = 1$. Finally, to better discriminate

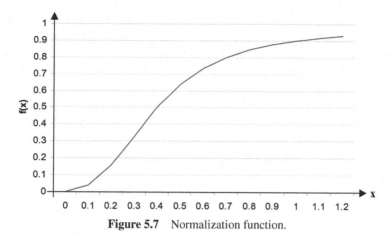

Figure 5.7 Normalization function.

between low- and high-affinity values, a quadratic form is introduced, leading to the following normalization function:

$$f(x) = \frac{a \tan\left(2\pi x^2\right)}{\frac{\pi}{2}}$$

Figure 5.7 shows the behavior of $f(x)$ in the interval [0, 2]. It should be noted like values on the *x-axis* greater than 1 (sum of high-value metrics) lead to limited effects on the meaning of the affinity.

The function $f(x)$, when applied to $W \cdot C_m^T$, provides affinity values that are directly comparable and therefore it can be used to select the best executors class for each functionality. Such information, together with others related to load and communication cost (Chapter 8), is an innovative building block of a cost function enabling the partitioning tool (Chapter 7) to efficiently explore the design space.

5.2.2 The Tool

In order to integrate in the *TOHSCA* environment the presented co-analysis methodology, and to validate the methodology itself, two equivalent tools have been developed to apply the methodology to two different languages: *OCCAM* and *C*.

In particular, due to the larger diffusion of *C* code (especially in the *DSP* field), and to provide a more meaningful methodology validation, the latter has been performed based on a *C* test suite. A proper tool has been developed and integrated with a C/C++ code analyzer (*GENOA*, [98]). The choice of GENOA is due to its great flexibility: in fact, it performs analysis on C/C++ code based on user-defined queries, expressed with a proprietary language. Such features allow for the execution of different analyses in a straightforward way allowing fast modifications on the features to consider during the methodology validation process.

Once the model has been validated, as shown in the following paragraph, an equivalent tool (the *OCCAM Analyzer*) has been developed in C++, based on the *aLICE* library, and integrated with the *OCCAM Parser* [10]. The output of such a tool is the affinity values for each system functionality; values that are provided to the system design exploration tools, like shown in Figure 5.1.

5.2.3 Validation

This paragraph presents the process adopted to validate the methodology and the metrics presented above. By means of the query-language provided by *GENOA,* it has been possible to extract from several *C* procedures the value of each metric and to evaluate the affinity for each procedure with respect to each executor class.

The adopted test suite is composed of 311 procedures; each of them represents a particular functionality to analyze. A subset of these procedures (i.e., 100, called *DSP-suite*) have been selected from applications oriented to digital signal processing and, therefore, they represent a valid sample of the main functionalities involved in these applications (e.g., *Fast Fourier Transform*, filterings, convolutions, etc.). The other procedures are representative of a general set that contains functionalities related to the field of coding (MD5, CRC, RSA, etc.), string manipulation (e.g., copy, compare, etc.), common operations (e.g., *sorting*), and parts of game implementations. Moreover, the general set contains also some hybrid (i.e., DSP/GPP) functionalities extracted from applications related to the management of digital signatures embedded into digital images.

During the validation process, the values of the metrics previously defined have been collected, and the affinity value of each functionality has been evaluated in the normalized form. Then, averages values have been evaluated for the procedures belonging to the whole suite, for the procedures belonging to the DSP-suite, and for the procedures not belonging to the DSP-suite (i.e., the *non-DSP-suite*). From Table 5.1, it is then possible to note an important feature: for the procedures in the DSP-suite, A_{DSP} in the DSP box is sensibly greater than the others affinity values and the A_{DSP} values evaluated in the non-DSP-suite. It is worth noting that the A_{GPP} has the large average along the

Table 5.1 Different affinity average values

	Average (whole test suite)	Average (*DSP-suite*)	Average (non-*DSP-suite*)
A_{GPP}	0.46	0.38	0.49
A_{DSP}	0.25	0.57	0.10
A_{HW}	0.27	0.28	0.27

whole set, to demonstrate the general purpose nature of the related executors class while the A_{HW} indicates in general (the three average values in Table 5.1 are nearly the same) those procedures that exploit some features associated with the *ASIC* executor class.

5.3 Conclusion

This chapter has addressed the definition of a set of metrics, providing quantitative information useful to make system-level decisions such as architectural selection and partitioning. The underlying idea is that the performance metrics of a final design can be related to the properties of the specification itself. Therefore, the core of this task involves identification and evaluation of functional and structural properties of specification, which could affect design performance on different architectural platforms.

In fact, in the co-design of heterogeneous multi-processor embedded systems, the choice of the processing elements that constitute the final architecture should be based not only on performance issues but also on the features of the tasks to be executed. The proposed metrics allow an effective exploration of the design space. Those metrics express the *affinity* of a functionality toward each possible processing element (*GPP, DSP*, and *ASIC/FPGA*), data that are then considered during the system design exploration step. To characterize the specification, the analysis process is constituted by different tasks: an architectural analysis of the existing processing elements, and the definition of a set of patterns able to identify subsets of the specification that could exploit the identified architectural features. Finally, a set of metrics has been defined in order to build a global metric providing straightforward and meaningful indications useful to make important architectural choices.

Finally, the chapter has presented the tools that have been integrated in the co-design environment and used to validate the methodology itself. The validation process has been described showing the experimental results obtained by applying the methodology to a meaningful test suite. Such results show the effectiveness of the proposed approach.

6

System-Level Co-Estimations

The first steps of the proposed flow (co-analysis and co-estimation, Figure 6.1) aim at obtaining as much information as possible about the system by analyzing the specification in a static manner. In particular, the goal of this chapter is to provide a methodology for the static estimation of the timing characterization of each system functionality, for both HW and SW implementations.

This chapter presents a strategy to efficiently analyze the timing characteristics of hardware/software embedded systems. The approach is suitable to enable design-space exploration during the early phases of the design process. It is based on a uniform modeling strategy for both hardware and software performance, to make possible the integration of the corresponding analysis model within the co-simulation engine (Chapter 8).

The range of hardware implementation strategies and the variability of the software design environments are taken into account at a high-level of abstraction using parametric models whose characteristics can be properly tuned.

For the software case, the proposed methodology concentrates on the essential aspects of the specification language and the common characteristics of assembly languages to derive an exact analytical model of the specification language structures. Deviations from this reference model, due to differences among microprocessors, compilers, and other environmental aspects, are accounted for with corrections derived on a statistical basis. The basic idea is to decompose the execution of a high-level statement into a sequence of simpler operations, at the granularity typical of the usual assembly languages, and then combine the contributions due to each elementary operation.

For the hardware case, the general approach is based on the same techniques adopted in the field of high-level synthesis to translate a high-level specification into a description at RTL level. The variability of hardware

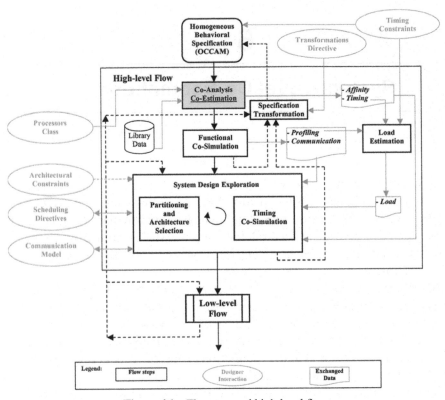

Figure 6.1   The proposed high-level flow.

resources and implementation strategies have been taken into account by modeling the execution time of hardware-bound statements in function of the available resource and different scheduling policies.

The resulting models characterize the static aspects of the specification while dynamic effects are accounted for by combining static data with actual profiling information. Experimental results, obtained after integrating the proposed methodology within the TOHSCA co-simulation engine, are discussed in detail.

The chapter is organized as follows: Section 6.1 describes system-level co-estimation issues, focusing on performance co-estimation, and Section 6.2 describes the proposed approach and shows its applicability to the OCCAM language. Finally, Section 6.3 draws some conclusions summarizing the potential benefits of the proposed approach.

6.1 Characterization

Estimation issues at high level of abstraction have been always considered carefully because of the relevant impacts that an early knowledge, as accurate as possible, of the features of the system being designed could have on the quality of the system itself, allowing a reduction of design errors and design times.

In the field of co-design, the co-estimation (i.e., concurrent estimations about hardware and software implementations) task has been focused on the main aspects of the design: power dissipation, cost (i.e., area), and performance.

Some representative attempts to consider cost and power dissipation at high-level of abstraction are shown in [10, 94, 97], while the aspect considered in this chapter, the performance estimation, is discussed in the following.

6.1.1 Performance Estimation

High-level performance estimation has been one of the more relevant research fields because of the possibility of early checking the meet of the timing constraints, a fundamental goal in the field of real-time systems.

Any analysis of the timing characteristics typically requires to address lower level (*cycle-true*) representations, which typically imply to deal with *RTL* descriptions for hardware components or assembly code running on microprocessor hardware models or instruction set simulators for the software components [56]. The timing estimation problem is thus afforded either by translating the description at system level into a finer grain, where each component of the system is accurately detailed and results are back-annotated, or by associating with the system-level components some coarse grain information.

The former approach [57, 58, 59] is characterized by good estimation accuracy. Nevertheless, it is time consuming and suffers some drawbacks such as the strong sensitivity to the considered design environment (e.g., the compiler) and architectural parameters, the need of cooperation among different analysis tools, and the decoupling of the estimation phase with respect to the *what-if* analysis loop for hardware/software partitioning. Such drawbacks make it inconvenient to be used in the early stages of the design process.

The latter approach [60, 61], working at higher level, is fairly independent of the technology and fast, frequently in detriment of accuracy. In fact, such a class of estimation methodologies is computed online, within the inner loop of the hardware/software analysis, but it does not take significantly into account the differences among target architectures.

6.2 The Proposed Approach

The proposed approach, relatively to those previously described, is a *meet in the middle* methodology, where the estimation is performed online at system-level, based on information coming from a performance model considering low-level characteristics of the possible (hardware or software) executors. The adopted timing estimation methodology consists in evaluating, at run-time, the timing behavior of each process involved in the system-level simulation. In such a way, it is easier to take into account unpredictable data-dependent conditions such as branches, loops, etc. The accuracy is good enough to take most of the decisions driving the hardware/software partitioning and behavioral analysis of the system. Furthermore, its parametric structure makes it a flexible approach with respect to the target technologies (Instruction Sets, clock period, etc.) and the development environment (compiler, synthesis tools, etc.). Even if the proposed methodology is general to be used in other environments, as a validation vehicle we considered the TOSCA co-design framework [4, 9, 11, 62].

This section is organized as follows: Section 6.2.1 formally introduces the problem, presents the methodology used for the estimation of the execution times detailing the timing model on which it is based on Section 6.2.1, and Section 6.2.2 shows the application of the methodology to the OCCAM language. Next, Section 6.2.3 provides some details about the tools developed for the estimation and about the integration of such tool in the TOSCA environment while Section 6.2.4 presents the results obtained by applying the proposed methodology to some simple, though significant, algorithms coded in OCCAM.

6.2.1 Model and Methodology

This paragraph formally introduces the problem of timing estimation, and presents the methodology used for the estimation of the execution times of the system-level specification in the software and hardware domains, detailing the timing model on which it is based on.

Problem definition

The goal of a high-level timing estimation model is to determine the execution time of a given system specification avoiding the actual compilation and execution (i.e., synthesis and simulation, for the hardware). To this purpose, a thorough analysis of the characteristics of a given specification language (or set of languages) is necessary. In the present work, the focus is on imperative, non *object-oriented* languages such as *C, FORTRAN, Pascal, OCCAM2,* and others. Moreover, the proposed approach allows the user to take into account arbitrarily wide classes of microprocessors by using a microprocessor conceptual description. Such a description includes the essential characterization of the instruction set of the processors taken into account. It is worth noting that, the wider is the microprocessors class, the less accurate is the result obtained but the faster is the exploration of the solution space.

The main problem consists in determining an estimate of the basic execution time of each different source-level statement. These estimates are based on a static model and are dynamically combined, according to the source code hierarchy, during the simulation phase (Chapter 8).

As an example consider the code portion reported in Figure 6.2(a).

The static models are applied to the blocks shown, along with their hierarchy, in Figure 6.2(b). The picture shows how, for example, the *while* statement is composed of a portion strictly related to the looping construct (*while* (...)...), a portion isolating the condition ($n < 10$) and a portion constituting the body of the loop (y [n++] = 0). When the source code is simulated with typical input data set, each timing contribution is accounted for according to the actual number of times the corresponding hierarchical block is executed. Figure 6.2(c) shows the possible execution paths and reports the number of executions of the blocks assuming that $a + b > 0$ and $n = 4$.

The overall methodology is thus static, constructive, hierarchical, and *branch-aware*. The static aspect is derived from the off-line evaluation of the time of each statement (but profiling is needed for the estimation of the execution time of a whole specification part). Such time estimation considers

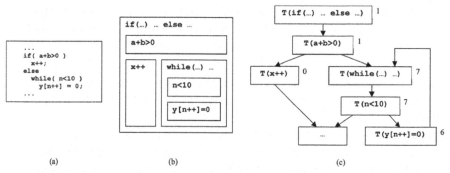

(a) (b) (c)

Figure 6.2 Sample code decomposition: (a) Source, (b) hierarchy, and (c) execution paths.

the essential characteristics of each statement by adding up the contribution of its different portions that are extracted by hierarchically analyzing the statement. The branch-awareness is obtained by producing, when necessary, pairs of values corresponding to the conditional statements class (*if, while*) in order to allow the simulator to adapt the timing characterization of such constructs to the data.

Methodology

A generic source program is composed of *statements* that can be classified into *operational statements*, such as arithmetic expressions and assignments and *control statements*, such as conditionals, loops, and function calls. On the other hand, the assembly program resulting from compilation of the source code is composed of *instructions* that are characterized by the *operation* they perform and the *addressing modes* used to access operands.

Operational statements are composed of operations (arithmetic, logic, relational, etc.) and operands (constants, scalars, vectors structured data types, etc.). Their translation is thus affected by the operations and addressing modes available in the target assembly.

Control statements, on the other hand, are typically translated into assembly code on the base of templates that use a small subset of instructions of the target language.

While the assembly operations and the translation templates are approximately independent of the target processor and their execution times can be easily estimated, the combined effects of the availability of certain addressing modes, the number of general purpose registers, and the optimization policies of the compiler introduce a high degree of variability.

Operational statements model

An effective statement is, in general, an assignment of the result of an expression to a variable and involves a number of simple operations such as sums, multiplications, index calculation and so on. The model of such a statement can be hierarchically built based on the models of the single operations carried out. In the best case, an arithmetic operation simply requires executing an instruction on one or two registers and storing the result again in a register.

Denoting with *alu* the generic arithmetic operation, the best-case execution time is simply $T_{\min}(\text{alu})$. In the worst case, the same operation requires loading n data from memory into registers, performing the operation itself, and then storing the result back to memory. The execution time is thus $n \cdot T_{\max}(\text{load}) + T_{\max}(\text{alu}) + T_{\max}(\text{store})$.

Control statements model

A control statement can be modeled referring to its *compilation template*. In general, its model is the sum of the times of the basic operations it uses. In the worst case, the times $T_{\max}(op)$ must be used while in the best case the times $T_{\min}(op)$ have to be considered.

It is worth noting that the compilation template is, in general, different in best-case and worst-case conditions.

Time model

A key point of the methodology concerns the adopted time model. Without loss of generality, we consider three types of time: software time, hardware time, and communication time. In order to make the simulation environment as flexible as possible, the concept of time has been abstracted, according to the considered types by using a dimensionless factor typically adopted for the software environment: the *CPI* (*Clock cycles Per Instructions*). The CPI, furthermore, averages all aspects depending on data characteristics. The CPI, thus, represents a tradeoff between accuracy and simplicity. Although the CPI concept has been introduced for the software partition only, the extension to other partitions is rather simple. In fact, any operation, independent of the partition it is assigned, needs time to be executed. By inheriting the same set of concepts from software, the CPI for a given operation is obtained as the ratio between the time needed to carry out the operation and the clock period of its partition. As far as communication is concerned, the only operation involved is equivalent to a data transfer. In this case, the time required to copy a datum depends on the protocol, on the data width W_{data} and bus width W_{bus} so that the CPI for the communication can be expressed as:

$$\text{cpi}_{\text{comm}} = \left\lceil \frac{W_{\text{data}}}{W_{\text{bus}}} \right\rceil \times g_1 + g_2$$

where g_2 only depends on the adopted transmission protocol, while g_1 accounts for the difference of the hardware and software clocks, and is mostly influenced by the response time of the slowest (hardware or software) partition. The communication time is not explicitly accounted for in the high-level static model of the statements but it is embedded into the simulation environment (Chapter 8) and can be configured according to the user needs.

For software programs, it is customary to define time in terms of CPIs. According to the definition of CPI, the actual time can be expressed as

$$T_{\gamma,\text{sw}} = \text{cpi}_{\text{SW}}(\gamma) \cdot T_{\text{CK},\text{sw}}$$

where $\text{cpi}_{\text{SW}}(\gamma)$ denotes the number of clock cycles required for the execution of the instruction γ and $T_{\text{CK,SW}}$ is the clock of the software partition, i.e., the clock period of the target microprocessor.

The concept of CPI has been extended and adapted to express the timing of hardware components as well. The instruction γ, implemented in hardware, is a generic operator whose execution time is mainly a function of its architecture (e.g., an *adder* can be implemented as *ripple carry, carry look-ahead* [63]), the width W_{data} of the operands and the *unit delay* Δ:

$$T_{\gamma,\text{HW}} = f(\Delta, \text{architecture}, W_{\text{data}})$$

The number of clock cycles to complete the execution of operation γ, i.e., the CPI for the hardware component, is given by

$$\text{cpi}_{\text{HW}}(\gamma) = \left\lceil \frac{T_{\gamma,\text{HW}}}{T_{\text{CK,HW}}} \right\rceil = \left\lceil \frac{f(\Delta, \text{architecture}, w)}{T_{\text{CK,HW}}} \right\rceil$$

According to the definitions of $\text{cpi}_{\text{SW}}(\gamma)$ and $\text{cpi}_{\text{HW}}(\gamma)$, the number of clock cycles necessary to execute a given operation can be compared easily. Nevertheless, the corresponding actual time still depends on the periods of the two clocks. To hide this dependence into a uniform model and to simplify the notations used throughout this chapter, a reference clock period must be defined and both measures referred to this clock. Let $T_{\text{CK}} = T_{\text{CK,SW}}$ be the reference clock period and T_{γ} the execution time of the operation γ, regardless of the partition to which the operation belongs. The CPI, being the ratio of an actual execution time and a reference time, can be thus redefined as:

$$\text{cpi}(\gamma) = \frac{T_{\gamma}}{T_{\text{CK}}} = \frac{T_{\gamma}}{T_{\text{CK,SW}}} = \text{cpi}_{\text{SW}}(\gamma)$$

for the software partition and as:

$$\text{cpi}(\gamma) = \frac{T_{\gamma}}{T_{\text{CK}}} = \frac{T_{\gamma}}{T_{\text{CK,HW}}} \cdot \frac{T_{\text{CK,HW}}}{T_{\text{CK,SW}}} = \text{cpi}_{\text{HW}}(\gamma) \cdot \frac{T_{\text{CK,HW}}}{T_{\text{CK,SW}}} = \text{cpi}_{\text{HW}}(\gamma) \cdot \Phi$$

for the hardware one. The ratio $\Phi = T_{\text{CK,HW}}/T_{\text{CK,SW}}$ is the factor that allows a comparison between hardware and software CPIs. In the rest of the chapter, the function $\text{cpi}(\gamma)$ will be used thus both for hardware and for software operations.

Software timing model

A given source code S is a list of statements γ_i, i.e., $S = \{\gamma_1, \gamma_2, \dots\}$. Each statement expresses a high-level operation that will be implemented as a suitable sequence of assembly instructions, according to a translation template or a compilation technique. As an example, let us consider two different statements: a looping construct such as *while* or *for*, and an arithmetic expression. The former can be translated according to a template scheme since its syntax is always the same. For the expression, the situation is quite different since, though the syntax is fixed, an expression can grow arbitrarily large and complex and thus no templates can be envisaged but rather each expression must be processed according to the structure of the corresponding syntax tree.

The original code can thus be modeled as a sequence of assembly instructions each composed of an *operation code* and a certain number of *operands* of different types. The operation code is strictly related to the task that must be performed to realize the desired high-level functionality and thus it is fixed. The number of operands supported by the assembly language is a characteristic of the instruction-set and is fixed too. To model a wide range of assembly languages, instruction-sets with 1-, 2-, and 3-operands instructions have been accounted for. The choice of a specific target microprocessor will determine which of the three cases to consider.

Owing to the reasons discussed above, the number of operands of the instruction set is fixed once the target processor has been selected and predictions on which operation codes will be used can be made based on the knowledge of the most popular compilation techniques and translation templates.

The addressing modes of the operands of the assembly instructions, on the other hand, are extremely hard to predict. The addressing modes used depend on a number of factors such as the class of the target processor (e.g., RISC/CISC), its architecture (e.g., the number of general purpose registers), and the compiler (e.g., optimization techniques).

To cope with this problem, the following strategy has been adopted. A generic instruction using complex addressing modes can always be decomposed in a suitable sequence of instructions using only simple addressing modes such as *immediate register direct* or *register indirect*. For example, consider the instruction shown in Figure 6.3: the first operand uses the *indexed* addressing mode, the second uses the *register indirect with auto-increment* addressing mode, and the third is a *register direct*; this instruction can be substituted by the code on the right side that uses only simple addressing modes.

It is reasonable to assume that a processor providing complex addressing modes has a number of units specifically dedicated to their management. These units are optimized and thus, probably, their use requires a shorter time than

Original instruction	Expanded instruction

ADD [R0,R4], [R2+], R3	`ADD R0,R4, R5`
	`LOAD [R5], R6`
	`LOAD [R2], R7`
	`ADD R2, #1, R2`
	`ADD R6, R7, R3`

Figure 6.3 Sample mapping to target assembly language.

the execution of the corresponding sequence of instructions exploiting simple addressing modes only. This assumption is in fact confirmed by the timing figures reported in the *programmer's manuals* of many microprocessors. The timing of the expanded instruction is always an overestimate of the actual timing.

Thanks to this observation, it is possible to translate an arbitrary source code into an assembly program using simple addressing mode instructions only, always resulting in a solution that overestimates the actual execution time.

Using a limited subset of instructions leads to an interesting generalization thanks to the fact that the basic operations that can be executed are roughly the same over a wide range of general-purpose processors. This concept is formalized in the following.

Let P_i be a generic microprocessor and IS_i its instruction set. Let then $P = \{P_1, P_2, \ldots, P_p\}$ be a set of p processors supporting instructions with the same maximum number of operands (typically one, two or three). The generic instruction set IS_i can be partitioned into a fixed number c of predefined *instruction classes* $IC_{i,j}$ performing similar operations, such as data transfer, load/store, branch, etc. The instruction classes must satisfy these relations:

$$\begin{cases} IS_i = \bigcup_{j=1}^c IC_{i,j} \\ IC_{i,j1} \cap IC_{i,j2} = \emptyset \quad \forall j_1, j_2 \in [1;c] \end{cases}$$

Instruction sets of different processors may significantly differ: for this reason, a specific processor may have one or more empty instruction classes. Two instructions $I_1 \in IS_1$ and $I_2 \in IS_2$ belonging to different instruction sets are said to be *compatible* if and only if:

$$\exists j \mid (I_1 \in IC_{1,j}) \wedge (I_2 \in IC_{2,j})$$

Considering all the p processors in P and their instruction sets IS_i, it is possible to define a number k of *compatible instruction classes* satisfying the following relation:

$$\text{CIC}_j = \begin{cases} \emptyset & \exists i | \text{IC}_{i,j} = \emptyset \\ \bigcup_{i=1}^{p} \text{IC}_{i,j} & \text{otherwise} \end{cases}$$

These new instruction classes collect all the instructions of different processors that are compatible in the sense that all the instructions in the same class perform equivalent operations. The union of all CIC_j classes can be thought of as a generic instruction set denoted as *KIS* or *Kernel Instruction Set*.

Let $\text{CIC}_j = \{I_{j,1}, I_{j,2}, \dots\}$ be the j-th compatible instruction class and N_j its cardinality. We can determine two instructions $I_{U,j}$ and $I_{L,j}$ in each compatible instruction class CIC_j such that their CPIs are maximum and minimum, respectively:

$$\begin{cases} I_{U,j} = \max_{n=1\ldots N_j} \text{cpi}(I_{j,n}) \\ I_{L,j} = \min_{n=1\ldots N_j} \text{cpi}(I_{j,n}) \end{cases}$$

The two instructions $I_{U,j}$ and $I_{L,j}$ represent the bounding cases for the j-th instruction class. Consider now a generic instruction I executed in $\text{cpi}(I)$ clock cycles. If I belongs to the j-th compatible instruction class then an upper-bound to its execution times is $\text{cpi}(I_{U,j})$ and, similarly, a lower-bound is $\text{cpi}(I_{L,j})$. If I does not belong to any of the compatible instruction classes, then there exists no single instruction in the compatible instruction set that can perform the same operation. Its functionality must thus be obtained by combining more than one instruction in *KIS*. The upper and lower bounds for an instruction $I \in \text{KIS}_1$ can thus be formally defined introducing the two functions:

$$\begin{cases} \text{cpi}_{\max}(I) = \text{cpi}(I_{U,j}) | I \in \text{CIC}_j \\ \text{cpi}_{\min}(I) = \text{cpi}(I_{L,j}) | I \in \text{CIC}_j \end{cases}$$

By using the instructions of the Kernel Instruction Set, it is thus possible to generalize the translation templates over multiple microprocessors and to account for the behavior of different compilers. However, as pointed out earlier, the uncertainty in predictions also comes from the architectural characteristics of the target processor. One of the most relevant parameters to consider is the number of available general-purpose registers. It is well known that the larger the number of registers, the smaller the number of spills that are necessary. Since spilling involves a memory access, which is extremely more time consuming than register access, the number of registers dramatically influences performance.

To deal with such variability, two theoretical limiting cases have been defined: an architecture with only two general-purpose registers and an architecture with an infinite number of registers. In the former case, a generic instruction requires loading the source data, performing the operation and storing back to memory the result, i.e., the *spilling* is maximized. In the latter case, all data can be thought as being pre-loaded into the general purpose registers and thus each instruction is executed by performing the operation directly on the registers, and no spilling at all is required.

For a more precise definition, it is worth introducing the two functions $cpp_{max}(\gamma)$ and $cpp_{min}(\gamma)$, referred to as the maximum and minimum *Clock cycles Per Process*.

These two functions combine the concepts of minimum and maximum CPI with the compilation issues related to the number of available registers. When a specific microprocessor is considered, the result of the compilation of a statement γ produces an assembly code falling in the range delimited by the two bounding cases just defined. Consequently, the timing in terms of CPIs of the considered statement lies in the interval $[cpp_{min}(\gamma); cpp_{max}(\gamma)]$. An estimate of the actual timing can be expressed as:

$$cpp_{est}(\gamma) = cpp_{min}(\gamma)^{\alpha} \cdot cpp_{max}(\gamma)^{(1-\alpha)}$$

where α is a parameter in the range $[0;1]$, accounting for the deviations from the two bounds. The proper value has to be determined by benchmarking the software compilation environment, through the analysis of different code segments, each producing a *local* α_j. Hence, the *global* value α used to represent a given microprocessor-compiler couple is

$$\alpha = \frac{1}{N_\alpha} \sum_{j=1}^{N_\alpha} \alpha_j$$

where the local parameters α_j have been determined, for each source code, in order to minimize the square error:

$$\varepsilon_j^2 = \left[cpp_{est}(\gamma) - cpp_{min}(\gamma)^{\alpha} \cdot cpp_{max}(\gamma)^{(1-\alpha)} \right]^2$$

This procedure, applied to the statement models described in the following sections, has been used to tune the overall performance estimation model.

Hardware timing model

Similarly to the software case, the variability of hardware resources and implementation strategies have been taken into account by modeling the execution time of hardware-bound statements, through the following relation:

$$cpp_{est}(\gamma) = cpp_{min}(\gamma)^{\beta} \cdot cpp_{max}(\gamma)^{(1-\beta)}$$

with the parameter β in the range [0;1]. Again, $\text{cpp}_{\min}(\gamma)$ and $\text{cpp}_{\max}(\gamma)$ represent the bounding cases for the execution time of each statement γ. The best case, $\text{cpp}_{\min}(\gamma)$, corresponds to an *ASAP* scheduling [64] with no bound on the number of functional resources. For the sake of completeness, note that such a value may slightly differ from the global minimum, since it does not take into account the effect of possible inter-statement optimizations. The worst case, $\text{cpp}_{\max}(\gamma)$, corresponds to the presence of a single functional resource per type, i.e., a purely serial implementation of the statement. The parameter β depends on both the number of resources and the scheduling policy. The proper value has to be determined by benchmarking the hardware synthesis environment, through the analysis of different code segments, each producing a local β_i, similarly to the procedure adopted for the software case.

6.2.2 Application of the model to OCCAM2

The theoretical model described in the previous paragraph has been applied to the OCCAM2 language and the software tools necessary to tune and validate the methodology have been developed. The following shows the details of such application for both software and hardware implementations, while the next paragraph gives some information about the tool.

Software

The time model has been applied to the main OCCAM statements considering a software executor. As highlighted in the previous section, the model is based on a compatible instruction set *CIS*, a set of mathematical expressions describing the constructs of the language and a tuning strategy. The defined compatible instruction classes set is shown in Table 6.1.

For each considered statement, a detailed analysis is presented in the following.

Table 6.1 The defined set of instruction classes

CIS	Operations
move	Data transfer from register or immediate to register
load store	Load from memory to register store register to memory
add sub and ...	Add registers Subtract registers Bitwise And on registers ...
cmp	Compare registers
jmp	Conditional or uncoditional jump and call

Variables and channels

The subset of the OCCAM2 language supported within the TOSCA environment supports *scalar* and *one-dimensional arrays*. Accessing a variable involves reading or writing directly from/to memory while the same operations on a channel require system calls in order to guarantee the semantics of the rendezvous synchronization protocol.

In the best case, all variables are already loaded into the infinitely many general-purpose registers available, so an access requires no additional operation with respect to those necessary for the actual processing. In worst-case conditions, a scalar variable access has the cost of a load operation. Accessing an element of a vector variable requires the index computation, i.e., evaluating the expression used as index, multiplying the index to the unit size of the variable of the specific type and, finally, accessing the element.

Channel access cannot be performed directly but rather by means of some suitable *system calls* that implement the rendezvous protocol, as described further in this paragraph.

Assignments

An assignment statement modifies the content of a variable. Note that the hierarchical nature of the proposed model allows us to concentrate on the assignment operation alone, considering the cost of the expression evaluation as an independent contribution. Expression execution time is, in fact, estimated separately (see below) and then combined with the cost of the assignment. The following reports minimum and maximum CPIs for the three types of possible assignments, corresponding to three different processes γ.

- *var = var*

   In the best case, a value stored in a register is transferred to another one while, in the worst case, a value stored in memory is copied to another memory location.

- *var = const*

   In the best case, a pre-loaded constant is copied in a register while, in the worst-case a constant value is moved to a specified memory location.

- *var = expr*

   The best case depends on the maximum number of operands supported by the target language: for two- or three-operands instructions, the expression evaluation can be translated in such a way that the result is directly stored in the target register while, for one-operand instructions the assignment requires an explicit move from the accumulator to the target register. In the worst case, regardless of the number of operands available, the result of the expression needs to be explicitly stored in the target memory location.

The timing models corresponding to these three cases are summarized in Table 6.2.

Arithmetic expressions

An arithmetic expression is a combination of symbols, operators, constants, and parentheses. Expressions follow the conventional rules of algebra. An expression can be easily represented by a *DAG* (*Direct Acyclic Graph*) such as that of Figure 6.4(a), where each internal node (V_I) and each leaf node (V_O) represent an operator and an operand, respectively. In the considered model, the set of internal nodes, V_I, has been decomposed into three subsets: V_{I0}, V_{I1}, and V_{I2}. The numerical indices denote the number of operands connected to the node. By indicating with $|V|$ the cardinality of V, and referring to Figure 6.4(b), yields $|V_{I0}|=2$, $|V_{I1}|=2$, and $|V_{I2}|=3$. In the best case, the CPIs depend on the instruction architecture. If three-operands instruction architecture is considered, the CPI is only related to both the number of algebraic operators and the overhead due to address calculation for array access.

If the instruction set contemplates algebraic instructions with less than three operands, further overhead has to be considered. In the case of two operand instructions, one of the source registers is always overwritten and needs thus to be saved. For this reason, one of the two variables connected to each internal node V_{I2} has to be moved in a temporary register to prevent its loss. In general, $|V_{I2}|$ variables have to be moved from their registers to temporary locations.

In the case of languages with one operand, the first operand is copied to an implicitly specified register (e.g., the accumulator) and, after the computation, is replaced by the result. The first operand to load in the implicit register is one of the two variables connected to each V_{I2} node. Note that the result must be moved from the accumulator and saved in a different location, when it is

Table 6.2 Models for software assignments

Process	cpp$_{min}$		cpp$_{max}$
	1 Operand	2, 3 Operands	1, 2, 3 Operands
var := var	cpi$_{min}$(move)	cpi$_{min}$(move)	cpi$_{max}$(load)+ cpi$_{max}$(store)
var := const	cpi$_{min}$(move)	cpi$_{min}$(move)	cpi$_{max}$(store)
var := expr	cpi$_{min}$(move)	0	cpi$_{max}$(store)

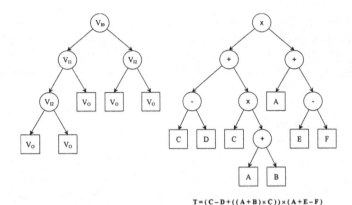

$$T = (C - D + ((A + B) \times C)) \times (A + E - F)$$

Figure 6.4   The considered model (a) and an example of algebraic expression graph (b).

not used by the next operation. In general, $|V_{I2}|$ variables and $|V_{I2}|$-1 partial results have to be moved. In the worst case (a target architecture providing two data registers only), the timing does not depend on the considered assembly language. As an example, consider the expression $(b + c) \times (d + e)$. The partial result corresponding to one of the sub-expressions $(b + c)$ or $(d + e)$ must be stored to make data registers available for the other sub-expression; then, the stored value must be re-loaded to compute the final result. In general, $|V_{I0}|$ results are involved in load-store operations. Tables 6.3 and 6.4 summarize the possible cases:

where:
$$W_{\mathrm{op,min}} = \sum_{i=1}^{i=|V|} \mathrm{cpi_{min}}(\mathrm{op}_i)$$
$$W_{\mathrm{op,max}} = \sum_{i=1}^{i=|V|} \mathrm{cpi_{max}}(\mathrm{op}_i)$$
$$W_{\mathrm{ld,max}} = (2 \cdot |V_{I2}| + |V_{I2}|) \cdot \mathrm{cpi_{max}}(\mathrm{load})$$

Table 6.3   Models for software arithmetic expressions (minimum values)

| Process | $\mathrm{cpp_{min}}$ | | |
	1 Operand	2 Operands	3 Operands
arit-expr	$W_{\mathrm{op,min}} + W_{\mathrm{arr,min}} +$ $(2|V_{I2}| - 1).\mathrm{cpi_{min}}(\mathrm{move})$	$W_{\mathrm{op,min}} + W_{\mathrm{arr,min}} +$ $|V_{I2}|.\mathrm{cpi_{min}}(\mathrm{move})$	$W_{\mathrm{op,min}} +$ $W_{\mathrm{arr,min}}$

Table 6.4   Models for software arithmetic expressions (maximum values)

| Process | $\mathrm{cpp_{max}}$ |
	1, 2, 3 Operands
arit-expr	$W_{\mathrm{op,max}} + W_{\mathrm{arr,min}} + W_{\mathrm{ld,max}} + |V_{I0}|.[\mathrm{cpi_{max}}(\mathrm{store}) + \mathrm{cpi_{max}}(\mathrm{load})]$

and $W_{arr,min}$ and $W_{arr,max}$ denote the overhead for accessing an array (i.e., to calculate the index value) in the best and worst cases, respectively.

Logical expressions

A logical expression is a combination of symbols, logical operations (AND, OR, and NOT), comparison operators (=, <>, <, >, >=, <=), constant, and parentheses. Logical expressions produce Boolean values. A logical expression can be modeled by an *OBDD (Ordered Binary Decision Diagram* [65]), representing a set of binary-valued decisions, culminating in an overall decision that can be either TRUE or FALSE. In the OBDD, one node can be a Boolean variable, i.e., a test applied to two variables of the same type or a constant. The nodes are ordered so that each level corresponds to a single variable.

It is worth noting that the performance is related to the graph shortest path that, in turn, depends on the local functions ordering. To overcome the computational effort of addressing all possible cases, justified by the goal of a fast performance evaluation, two different approaches have been adopted. The first approach is based on the two following assumptions.

The shortest path length is 1, while the longest path one equals the number of local functions (L). An analysis considering a set of examples has justified this first assumption: a typical logical expression is composed of a limited number of local functions (generally less then 4) and, usually, the shortest path evaluates 1.

To preserve the functionality (side-effects) of a given logical expression, shortcuts should be avoided and all the local functions should be pre-computed. This assumption allows a modular composition of the different contributions. Therefore, minimum and maximum CPIs are given by the relation shown in Table 6.5.

The second approach considers always the worst cases (i.e., path length always equal to L) and is particularly significant when dealing with hard real-time applications. The costs are given in Table 6.6.

Table 6.5 Models for software logic expressions (best case)

Process	cpp_{min}		cpp_{max}
	1 Operand	2, 3 Operands	1, 2, 3 Operands
logic-expr	$cpi_{min}(cmp) +$ $cpi_{min}(move) +$ $cpi_{min}(jmp)$	$cpi_{min}(cmp) +$ $cpi_{min}(jmp)$	$W_{ld,max} + L[cpi_{max}(cmp)$ $+ cpi_{max}(jmp)]$

Table 6.6   Models for software logic expressions (worst case)

Process	cpp_{min}		cpp_{max}
	1 Operand	2, 3 Operands	1, 2, 3 Operands
logic-expr	$L[cpi_{min}(cmp) +$ $cpi_{min}(move) +$ $cpi_{min}(jmp)]$	$L[cpi_{min}(cmp)$ $+ cpi_{min}(jmp)]$	$W_{ld,max} + L[cpi_{max}(cmp)$ $+ cpi_{max}(jmp)]$

Statement IF

The syntax of the conditional statement IF, shown in Figure 6.5, is structured in condition-process couples and, from a semantical point of view, resembles that of the *switch* statement of the *C* language.

The last condition-process couple has the purpose of avoiding a deadlock. According to the semantics of the IF statement, in fact, the conditions are evaluated in sequence and the process corresponding to a true condition is executed. When no condition yields a true value, the statement is equivalent to the STOP statement, whose purpose is to force a deadlock.

Executing the statement requires first evaluating the logical expression <*cond1*> and storing its result to a register (best-case) or to a memory location (worst-case). The result must then be compared with a constant value, either TRUE or FALSE, and a decision taken. For the following discussion, refer to the block diagram of Figure 6.6.

If the first condition is TRUE, then the code corresponding to its process, which is likely to be written immediately after the condition evaluation code, is executed and a branch to the first statement following the IF is performed. On the other hand, if the first condition is FALSE, a jump to the section of assembly code devoted to the evaluation of the second condition must be performed. After that, the same operation performed for the first branch is required to conclude the execution of the second branch.

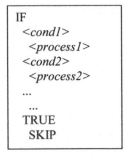

Figure 6.5   Syntax of the IF statement.

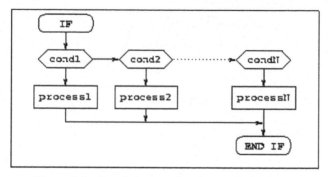

Figure 6.6 Translation template for the IF statement.

The overall cost depends thus from the number of branches, and in particular:

- In the best-case, the cost of each branch is solely that of a comparison and a branch, either to the next condition code or from the end of the process code to the end of the IF statement. In the optimal situation, the first condition is TRUE and no other conditions need to be evaluated. The cost is thus independent of the overall number of branches.
- In the worst-case, the cost of each branch is that of loading the result of the expression in a register, comparing it with a reference value and then jumping to either to the next condition or from the end of the process to the end of the IF statement.

Table 6.7 summarizes the mathematical expressions of the model that to be used dynamically. This means that the cost expressed by the relations of the table is related to the execution of one of the branches. Data dependencies must be accounted for by means of simulation only while must be neglected in this static analysis.

Statement WHILE

The WHILE statement is the only form of looping supported by TOSCA subset of the OCCAM 2 language. Its syntax, shown in Figure 6.7, is identical to that of many other languages, such as C, Pascal, or FORTRAN.

Table 6.7 Timing model for the IF control statement

	cpp_{min}	cpp_{max}
Process	1, 2, 3 Operands	1, 2, 3 Operands
IF	$cpi_{min}(cmp) + cpi_{min}(jmp)$	$cpi_{max}(load) +$ $cpi_{max}(cmp) + cpi_{max}(jmp)$

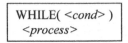

Figure 6.7   Syntax of the WHILE statement.

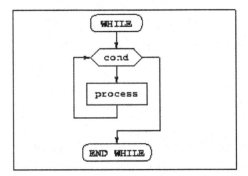

Figure 6.8   Translation template for the WHILE statement.

Table 6.8   Timing model for the WHILE

	cpp$_{min}$	cpp$_{max}$
Process	1, 2, 3 Operands	1, 2, 3 Operands
WHILE	cpi$_{min}$(cmp) + cpi$_{min}$(jmp)	cpi$_{max}$(load) + cpi$_{max}$(cmp) + cpi$_{max}$(jmp)

The body of the loop can be a simple statement, such as an assignment, or a compound statement enclosed by container process. A possible translation template for this container process is reported in Figure 6.8. This template is actually used by most compilers.

In the best case, the result of the expression, used as condition, is already stored in a register and thus can be directly compared against the constant used to represent the logical truth. Then, each iteration of the loop always requires a jump: back to the code for the condition evaluation, or to the first instruction out of the body of the loop.

In the worst-case, the number of jumps is again one, while the test of the condition requires loading the result of the expression into a register before performing the actual comparison.

The timing models for the WHILE process are shown in Table 6.8.

Communication processes

The TOSCA subset of the OCCAM2 language provides two communication primitives: the input process, the output process, respectively denoted by the symbols ?, !.

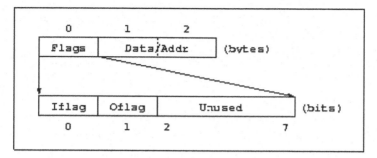

Figure 6.9 Data structure for channels implementation.

Communication takes place over unidirectional, blocking, *point-to-point* channels according to the rendezvous protocol. Each channel is thus bound to a writer and a reader and communication cannot occur unless both processes are ready to be executed.

This means that if a process attempts a write operation on a given channel, the operation does not take place until the corresponding reader is ready. Similarly, a reading process suspends until the corresponding writing process is ready. When both processes bound to a channel are ready, communication over the channel occurs.

The implementation of this protocol can be done in a number of ways, ranging from well-known mechanisms, such as semaphores, mail-boxes etc., to ad-hoc algorithms, studied in order to be efficient for the specific needs, with a consequent loss of generality.

The implementation of the rendezvous protocol used by the *OCCAM2-to-VIS compiler* uses a data structure, stored in a memory area with global scope, which contains the fields described in Figure 6.9.

The 1-bit flag *Iflag* (*Oflag*) is set to 1 when the input (output) process bound to the channel is ready while it is reset to 0 when the process is not ready. The 16-bit field *Data/Addr* contains the data to be written to the channel when an output process is invoked first or the address of the variable into which the data are to be read, when an input process is executed first.

In the following, the operations needed to carry out the communication over channels are described in detail for both the input and output processes.

Input process

The operations to execute an input process depend on the output flag:

- **Oflag = 1**

Table 6.9   Timing model for the input process

Process	cpp_{min} 1, 2, 3 Operands	cpp_{max} 1, 2, 3 Operands
chan ? var	$cpi_{min}(load) + cpi_{min}(cmp) +$ $cpi_{min}(jmp) + 2 \cdot cpi_{min}(store)$	$2 \cdot cpi_{max}(load) + cpi_{max}(cmp) +$ $cpi_{max}(jmp) + 2 \cdot cpi_{max}(store)$

A process is ready to communicate and thus the field *Data/Addr* contains the datum to be read into the variable whose address is known by the input process being executed; the datum is then copied and the two flags are reset to 0.

- **Oflag = 0**

No output process is ready to communicate and thus the input process must suspend. To do this, the flag *Iflag* is set to 1 and the field *Data/Addr* is set to the address of the variable specified as target in the input process.

The operations required are thus a comparison (for the *Oflag*), the corresponding jump, a data copy operation (either the datum or the variable address in the field *Data/Addr*), and a bit-set operation, which is supposed to be performed with a data copy instruction.

It is essential to note that the data structures designed to hold the status of channels are stored in a global memory area. For this reason, it is reasonable to assume that also in best-case conditions, *load* operations are necessary to access this memory instead of the usual *move* instructions.

The relations expressing the timings of the input process are summarized in Table 6.9.

Output process

The operations to execute an output process depend on the input flag:

- **Iflag=1**

A process is ready to communicate and thus the datum to be communicated is copied in memory at the address specified by the field *Data/Addr*. Then, both flags are reset and the value of the *Data/Addr*, which is left unchanged, becomes meaningless.

- **Iflag=0**

No input process is ready to communicate and thus the output process must suspend. To do this, the flag *Oflag* is set to 1 and the datum to be communicated is copied to the field *Data/Addr*.

The operations necessary to accomplish this process are summarized in Table 6.10.

Statement ALT

The arbiter process ALT is a combination of a number of input processes and an IF-like construct. The arbiter process supported and implemented by the TOSCA compiler includes the idea of priority and has the semantics of the OCCAM2 PRI ALT process. It follows the syntax described in Figure 6.10.

The guards *<guard1>* and *<guard2>*, etc., are couples of Boolean expressions and input processes. Their effect is to produce the execution of the corresponding process if and only if the input process can be executed (i.e., an output process is ready to write on a given channel) and the Boolean condition evaluates to TRUE. Once one of the branches is executed, the ALT process terminates. If none of the guards is satisfied, the process loops, waiting for a guard to become executable.

It is worth noting that when executing a normal input process, if no writers are ready, the input process must suspend itself while when executing the input process of a guard, the input process must not suspend.

The execution of this arbiter involves:

- a loop scanning all the guards in sequence;
- the evaluation of a logical expression; and
- the execution of an input process, with the modified semantics.

Table 6.10   Timing model for the output process

Process	cpp_{min} 1, 2, 3 Operands	cpp_{max} 1, 2, 3 Operands
chan ! var	$cpi_{min}(load) + cpi_{min}(cmp) +$ $cpi_{min}(jmp) + 2 \cdot cpi_{min}(store)$	$2 \cdot cpi_{max}(load) + cpi_{max}(cmp) +$ $cpi_{max}(jmp) + 2 \cdot cpi_{max}(store)$

```
ALT
  <guard1>
    <process1>
  <guard2>
    <process2>
  ...
```

Figure 6.10   Syntax of the ALT statement.

Let a guard have the syntax ($<channel>$? $<var>$) & $<cond>$: the input process, apart from its inherent execution cost, can be seen as a Boolean expression returning TRUE on execution and FALSE otherwise.

In this case, the cost of the guard itself is calculated adding three contributions: the cost of the condition *cpp(cond)*, the cost of the guarded-input process *cpp(ch?var)* including possible array-access overheads, and that of the logical AND.

The guarded-input process requires the same operation of the normal input when the communication actually takes place while only a comparison is needed if there are no waiting output processes. The costs are reported in Table 6.11.

The cost of the container ALT alone is similar to that of the IF process with the difference that the logical AND connecting the guarded-input and the condition requires an additional compare and an additional jump. In the worst-case, the variable operand of the compare operation must be loaded first. The complete expression of this cost is reported in Table 6.12, where, as for the IF statement, the costs refer to the execution of one of the branches. Data dependencies must be accounted for by means of simulation.

Hardware

Like for the software case, the time model has been applied to the same OCCAM statements considering now a dedicated hardware executor. For each statement, a detailed analysis in order to allow the application of the methodology for both the hardware and the software cases is presented.

The general approach is based on the same strategy adopted to translate OCCAM system-level models into a VHDL description at RTL level [10]. The focus is on the estimation of the number of CPIs needed in the best

Table 6.11   Timing model for the guarded-input process

Process	cpp_{min} 1, 2, 3 Operands	cpp_{max} 1, 2, 3 Operands
chan ? var (guarded)	$cpi_{min}(load) + cpi_{min}(cmp) + cpi_{min}(jmp)$	$2 \cdot cpi_{max}(load) + cpi_{max}(cmp) + cpi_{max}(jmp) + 2 \cdot cpi_{max}(store)$

Table 6.12   Timing model for the ALT arbiter

Process	cpp_{min} 1, 2, 3 Operands	cpp_{max} 1, 2, 3 Operands
ALT	$cpi_{min}(cmp) + 2cpi_{min}(jmp)$	$cpi_{max}(load) + cpi_{max}(cmp) + 2cpi_{max}(jmp)$

and worst cases analyzing the templates that correspond to each OCCAM statement. Such templates are combinatorial (e.g., for expressions evaluation) or sequential (e.g., for container processes) machines that implement the correct behavior.

Assignments

As stated before, the assignment statement modifies a variable. Three different situations can be identified:

- *var = var*

a value stored in a register is transferred in a different one;

- *var = const*

a constant value is copied in a register;

- *var = expr*

the expression result is loaded in a register. Since expression's computation and result storing can occur simultaneously, the CPI corresponding to this process is completely dominated by the expression computation.

The models, that in this case are constant, are reported in Table 6.13.

Expressions

In the best case, the execution time corresponds to a scheduling without resources limitation, i.e., the intrinsic process parallelism is maximized. In such a case, a good approximation of best performance can be achieved if the DAG corresponding to the expression is perfectly balanced and operators are all of the type of the fastest operator involved. The DAG depth can be expressed as the logarithm base 2 of the number of nodes plus one.

The worst case refers to an architecture where the computation is serial. By considering the DAG representation introduced before and allocating each operator on a different scheduling step, the total timing can be expressed as the sum of the timings of all the operators involved. These relations are summarized in Table 6.14.

Table 6.13 Models for hardware assignments

Process	cpp_{min}	cpp_{max}
var := var	1	1
var := const	1	1
var := expr	0	0

Table 6.14   Models for hardware arithmetic expressions

Process	cpp_{min}	cpp_{max}
expr	$\log_2(|V|+1) \cdot$ $\min\{cpi_{min}(op)\}$	$\Sigma_{op \in V}$ $cpi_{max}(op)$

Arithmetic operators

In order to produce a characterization as close as possible to the user needs, the most representative architectures of operators have been considered [63]. Table 6.15 reports the models used to evaluate the timing of their implementations.

Logical and relational operators

The relational operators $<$, $>$, and $=$ are considered like a subtraction and a check of some status bits (i.e., zero bit and sign bit) so their cost is the same as the minus operator plus a unit-delay for the check. For the logic operators have been supposed a tree of logic ports, which in depth depends on the word width. Table 6.16 summarizes the possible cases.

Statement SEQ

The template related to the *SEQ* process is a finite-state machine that allows the sequential execution of the children processes. Figure 6.11 represents a schematic view of a possible hardware implementation of a three-children *SEQ* process, highlighting the control signals.

When the input signal *Start* rises, it resets the counter and so the decoder output activates the corresponding sub-process. The input signals *Start* should be kept high by the *SEQ* father process while the output signal *Finish* is 0. When the first child terminates the execution, it raises the signal f_1, the counter

Table 6.15   Timing performance of the main hardware operators implementations

Operand	Architecture	$f(\Delta, \text{Architecture}, w)$
+/−	Ripple carry	$2 \cdot w \cdot \Delta$
	CCLA(n/p*p) p = BCLA dimension	$[10 + \log_p(1 + w \cdot (\frac{w}{4} - 1) \cdot (\frac{w}{4} + 1))] \cdot 11 \cdot \Delta$
*	Baugh Wooley	$4 \cdot w \cdot \Delta$
	Bisection	$4 \cdot w \cdot \Delta$
/	Restoring (Dean)	$(3 \cdot w^2 + 1) \cdot \Delta$
	Non Restoring (Guild)	$3 \cdot (w + 1)^2 \cdot \Delta$
	Non-restoring with 2-level CLA and carry-save	$(11 \cdot w + 12)^2 \cdot \Delta$
	Non-restoring with 2-level CLA and carry-save	$(9 \cdot w + 10)^2 \cdot \Delta$

Table 6.16 Timing performance of the hardware logical and relational operators implementations

Operand	Architecture	$f(\Delta, \mathrm{Architecture}, w)$
$<, >, =$	Ripple carry	$2 \cdot w \cdot \Delta + \Delta$
	CCLA(n/p*p) p = BCLA dimension	$[10 + \log_p(1 + w \cdot (\frac{w}{4} - 1) \cdot (\frac{w}{4} + 1))] \cdot 11 \cdot \Delta + \Delta$
$\cup, \cap, \oplus, !$	$\Delta \cdot \log_2(w)$	

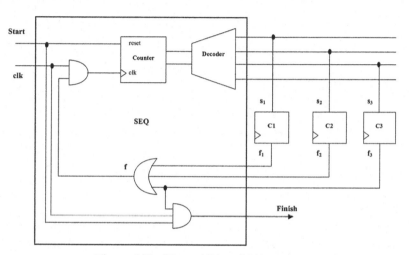

Figure 6.11 Three-children SEQ template.

is incremented and the next sub-process is activated. Since the signals f_i are synchronous, such activation requires one clock cycle. The whole process is repeated until the f_n signal rises the output signal *Finish*.

Therefore, the *SEQ* process requires one clock cycle for each child.

Statement PAR

The template for a *PAR* process is simpler than the one for the *SEQ*. In fact, a combinatorial network, represented in Figure 6.12, allows the concurrent execution of the children processes.

When the input signal *Start* is raised, the children processes are concurrently activated without additional clock cycles. The input signals *Start* should be kept high by the *SEQ* father process while the output signal *Finish* is 0. When all the children terminate the execution, all the signals f_i are high and so f switches to 1. At the following positive clock edge (the signals f_i are synchronous), the signal *Finish* raises.

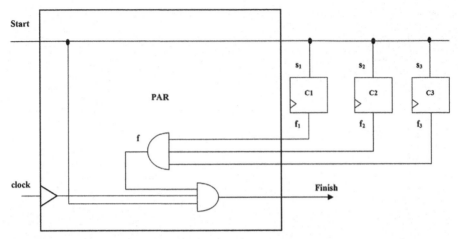

Figure 6.12   Three-children PAR template.

Therefore, the cost of a *PAR* process is of 1 clock cycle.

Statement IF

The template for an *IF* process is similar to the *SEQ* one, but this time each child is associated with a condition. A possible hardware implementation of an IF process with three children is given in Figure 6.13.

When the *Start* signal is raised, the decoder activates the first condition evaluation.

If the condition is true, the correspondent process is activated. When such process terminates, the signal f_i is raised and so, at the next positive clock edge (the signals f_i are synchronous), the signal *Finish* switches to 1. If a condition is false, the counter is incremented and another condition is evaluated.

Moreover, it should be noted that, in the *IF* process, it is always present the *TRUE-SKIP* condition in order to avoid deadlock. Such condition is represented by the line directly connected with the final *OR* port.

So, the cost of an *IF* process is of one clock cycle for each branch.

Statement WHILE

The *WHILE* process is similar to a *one-condition IF*, but this time the execution of the only child process is repeated while the condition is true and so to manage this statement a combinatorial network is sufficient. Figure 6.14 shows a schematic representation of a possible hardware implementation of the *WHILE* control statement.

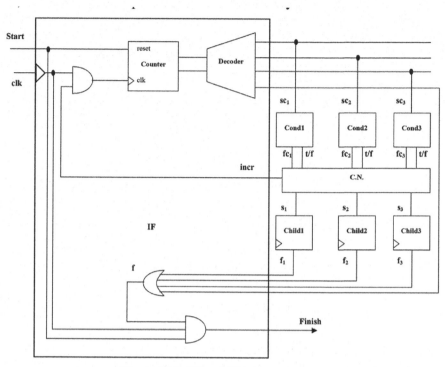

Figure 6.13 Three-children IF template.

When the input signal *Start* switches to 1 the condition is evaluated. When the evaluation process ends, *fc* is used to reset *sc* to 0 allowing the detection of a new signal edge.

If the condition is false, the output signal *Finish* is raised at the following positive clock edge and the process ends. If the condition is true, the child process is executed. When the child ends, *f* gives rise to another condition evaluation and the whole loop is repeated.

Therefore, the WHILE process introduces one additional clock cycle for each loop.

Communication processes

The hardware implementation of the rendezvous protocol used by OCCAM channels takes advantage of the fact that a channel is an *unidirectional, point-to-point* logical connection: in hardware, such a connection is mapped into a physical one, while for sw-hw (and hw-sw) communications, a bus interface among the hardware and the related processor is needed.

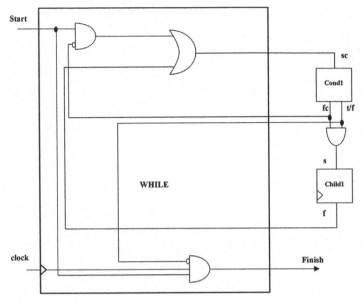

Figure 6.14   WHILE template.

The overall structure is represented in Figure 6.15.

Figure 6.16 shows a state-diagram that represents the behavior of a channel. A complete rendezvous is represented in Figure 6.17. In order to consider only the additional clock cycles due to the channel management, the case in which the input and output processes are ready at the same time is analyzed.

When *StartW* rises, a positive clock edge is needed in order to detect the signal edge allowing also writing data into the channel and changing the state from *DNP* to *DP*. In the following positive clock edge, *StartR* is already high

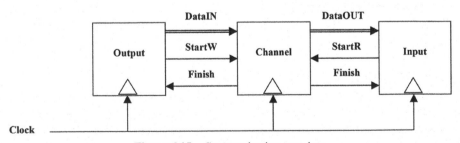

Figure 6.15   Communication template.

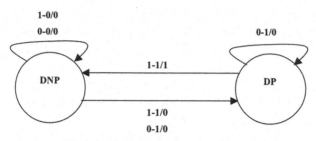

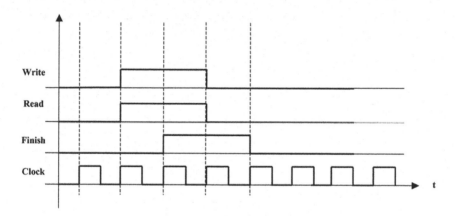

Figure 6.16 Channel state diagram.

Figure 6.17 A complete rendezvous.

and then the read operation can be performed: the output signal *Finish* raises, and the state changes to *DNP*. Finally, the high value of *Finish* allows the communication processes to fall *StartW* and *StartR*. At the following positive clock edge a new communication can be started.

Therefore, the overhead due to the channel management is of two clock cycles.

Statement ALT

The last control statement considered is the *ALT* one. It is important to note that the *ALT* process implemented in TO(H)SCA is in reality a *PRIORITY-ALT*. In fact, a fixed priority is assigned to the children by means of a combinatorial network. This is necessary in order to avoid the non-determinism of the original

ALT. Its template is a combinatorial network like the one shown in Figure 6.18 for a two-children case.

When the input signal *Start* raises, all the conditions are concurrently evaluated and the internal signal r switches to 0 in order to allow a new positive clock edge detection. The father process should keep *Start* high while *Finish* is 0.

If a condition is true, the related channel is checked in order to control the presence of a data (this is due without performing a real read but looking directly to the internal state S of the channel).

Thus, a combinatorial network allows selecting the child with the condition true, the channel ready, and the greater priority (fixed in the combinatorial network *C.N.*). The selected channel is now read really and the related child is executed. Its termination raises, in the following positive clock edge, the output signal *Finish*.

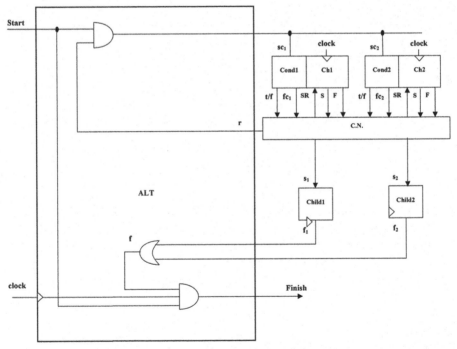

Figure 6.18   Two-children ALT template.

Instead, if no processes are executable, the internal signal r switches to 1 starting another conditions evaluation. In this way, the *ALT* process loops waiting for a child to execute.

Therefore, the overhead due to the *ALT* management is the one due to the *channels* management (two clock cycles) plus one clock cycle to synchronize the output signal *Finish*.

6.2.3 The Tool

The presented model has been integrated into the TOHSCA framework as a stand-alone *C++* tool: *SLET (System-level Estimation Tool)*. The tool is based on three components.

- **Kernel**

  The kernel is the executable portion of the tool and has the main purpose of loading and parsing an OCCAM2 source file, generating an internal representation. The data structure on which the internal model is based is a tree of C++ objects (see Section 2.2.3), each one representing an element of the language. The outcome is thus a decorated tree. The objects contain a number of methods that can be used to extract information about the OCCAM2 element they model. One of these methods has the purpose of calculating the estimated timing based on the high-level model previously presented. To do this, the algorithm visits the sub-tree originating at the current object and collects data such as the number of operators used, the number and type of variables, and so on. These data are stored in a suitable structure, which is then passed as an argument to the library function that actually performs the cost computation (see below).

- **CPIs File**

  This file is a library containing timing data on the assembly instruction classes. For each instruction class, the *cpimin* and *cpimax* values are given. A sample file is depicted in Table 6.17.

- **Timing-Model Library**

  This library collects the mathematical expressions to be used to compute the execution time of OCCAM2 processes. For each process, a function, named after the process, computes the timing costs based on the data found in the structure it receives as an argument. The library is compiled as a shared object in order to minimize the coupling of the data collection phase and the actual computation phase.

Table 6.17   Sample CPI file for the Intel i80486DX2 processor

# INTEL i80486		
Operands: 2		
# class	min	max
ADD	1	3
CMP	1	2
JMP	1	3
...	...	...
STORE	1	4

When the estimation tool is run, the dynamic library containing the model computation functions is opened and linked. Then, the CPI file corresponding to the desired target microprocessor is read and loaded into the data structure partly shown in Figure 6.19.

To clarify the behavior of the tool, consider the simple example of the assignment process $Y:=4*(A + B + 1)$. Its tree representation, based on the class library aLICE [10], which constitutes the core of all tools reading the OCCAM files, is shown in Figure 6.20.

The method *CAssignment::count*(), when invoked, calls its polymorphic implementations for the two elements *Left* and *Right* of its children list. This happens for all objects that are not leaves of the tree. When a leaf is reached, the method *count*() simply increments a suitable counter, which is part of the data structure passed to the method itself.

When this visit terminates, the structure returned in the instance of *CAssigment* by the method *count*() completely describes the whole sub-tree. Figure 6.21 shows a portion of the data structure.

This structure is then passed to the dynamically linked computation function *calc_assigment*(), passing the two structures *cnt* and *cpi*. The function

```
struct cpi
{
        int operands;
        int ADD[2];
        int CMP[2];
        int JMP[2];
        ...
        int STORE[2];
};
```

Figure 6.19   CPI and operator count data structure.

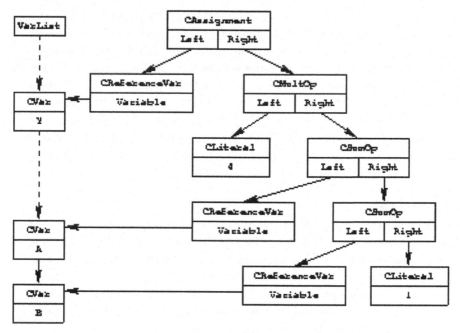

Figure 6.20 Tree model of the assignment $Y:=4*(A+B+1)$.

```
struct cnt
{
        int plus;          /* Additions    */
        int minus;         /*Subtraction  */
        ...
        int int_vars;      /*Integer vars */
        int float_vars;    /*Float vars   */
        ...
}
```

Figure 6.21 Operator and variable counts data structure.

embeds the mathematical model for the OCCAM2 statement under consideration and returns the minimum and maximum overall costs. These timings are then stored in a member of the *CAssigment* object.

6.2.4 Validation

The presented methodology has been applied to a number of benchmarks and some the results are described in the following. The models have been developed for four microprocessors (*Intel 80486DX2, Sparc V8, Arm Ltd. ARM7TDMI* in *Thumb mode* and *Motorola 68000*) and for the *LSI Logic 10k* technology. These models have been implemented and integrated in the TOHSCA co-design environment that has been used as the validation framework. The operating conditions of the experiments reported below assume a clock frequency of 33 MHz for the software partition and a system clock of 150 MHz for the hardware devices. The scaling factor Φ is thus 0.22. The results for the two benchmarks *GCD* (Euler's algorithm for greatest common divider calculation) and *BSORT* (Bubble sort algorithm) are presented in detail.

Example 1: GCD

The GCD algorithm applied to 16-bit integers has been specified in OCCAM and run through the estimation flow based on the tool SLET, yielding the static figures reported in Table 6.18. These data have been used for simulation using the system-level simulation tool OSTE (Chapter 8) and two sets of inputs, leading to the results summarized in Table 6.19. The actual execution time for the algorithm has been derived from simulation of the low-level models: assembly code, for the software case, and gate-level VHDL for the hardware case. The assembly code has been generated and simulated using the TOSCA tools while the gate-level model has been synthesized and simulated using a commercial tool [106].

The results show a relative error averagely below 8%. This represents surely an optimal trade-off between accuracy and estimation times. In fact, the static estimation of the execution times for hardware and software is extremely fast and, for files of a few tenths of lines, can be neglected.

Table 6.18   Static timing estimates for the GCD algorithm

Source	Estimated CPIs				
	i486	Sparc	ARM7	MC68000	LSI 10k
PROC gcd(INT a, INT b)	0.00	0.00	0.00	0.00	0.00
WHILE(a<>b)	18.13	5.95	7.23	16.25	1.39
IF	13.31	2.38	2.61	8.36	0.28
(a>b)	25.54	9.67	9.96	12.26	8.73
a := A − b	6.40	1.41	1.75	2.25	5.34
(TRUE)	0.00	0.00	0.00	0.00	0.00
a := b − a :	6.40	1.41	1.75	2.25	5.34
	0.00	0.00	0.00	0.00	0.00

Table 6.19 Dynamic timing estimates, measures and errors for the GCD algorithm

Input	Measure	i486	Sparc	ARM7	MC68000	LSI 10k
Set 1	Actual	48314	12273	16463	17472	15683
	Estimated	48505	11034	14740	19524	16985
	Error %	−0.40	10.09	10.47	−11.75	−8.30
Set 2	Actual	10114	2569	3446	3657	3283
	Estimated	10798	2938	3209	3446	3444
	Error %	−6.77	−14.36	6.90	5.77	−4.89

Table 6.20 Static timing estimates for the BSORT algorithm

	Estimated CPIs				
Source	i486	Sparc	ARM7	MC68000	LSI 10k
PROC bsort([10]	0.00	0.00	0.00	0.00	0.00
INT v) INT i,j,t:	0.00	0.00	0.00	0.00	0.00
SEQ	0.00	0.00	0.00	0.00	0.00
i:= 0	3.33	1.19	1.44	4.64	0.29
	18.13	5.95	7.23	16.25	1.40
WHILE(i<9)					
SEQ	0.00	0.00	0.00	0.00	0.00
j:= 0	3.33	1.19	1.30	3.31	0.29
WHILE (j<9−i)	12.12	7.14	8.40	16.75	3.18
SEQ	0.00	0.00	0.00	0.00	0.00
IF	13.31	2.38	2.61	8.36	0.29
(v[j+1]<v[j])	25.54	9.67	9.96	12.26	1.92
SEQ	0.00	0.00	0.00	0.00	0.00
t:=v[j]	3.33	2.39	3.09	3.06	0.29
v[j]:=v[j+1]	3.33	2.39	2.60	3.69	0.78
v[j+1]:=t	3.33	2.39	2.73	6.30	0.78
(TRUE)	0.00	0.00	0.00	0.00	0.00
SKIP	0.00	0.00	0.00	0.00	0.00
j:=j+1	6.40	1.41	1.75	2.25	1.17
i:=i+1	6.40	1.41	1.75	2.25	1.17
:	0.00	0.00	0.00	0.00	0.00

Example 2: Bubble sort

The same procedure has been applied to a slightly more complex example: the bubble sort algorithm. The details of the static execution times are given in Table 6.20, while the corresponding dynamic measures are reported in Table 6.21. With this benchmark also, two different input vectors with random entries have been used.

The second example gives rise to errors and considerations similar to the previous one, enforcing the validity of the presented estimation methodology.

Table 6.21   Dynamic timing estimates, measures and errors for the BSORT algorithm

Input	Measure	i486	Sparc	ARM7	MC68000	LSI 10k
	Actual	2958	1168	1322	2340	370
Set 1	Estimated	2716	1053	1364	2480	393
	Error %	8.19	9.88	−3.23	−6.01	6.21
	Actual	2070	818	925	1638	259
Set 2	Estimated	2096	918	946	1843	301
	Error %	−1.25	−12.23	−2.23	−12.53	16.21

6.3 Conclusion

This chapter addressed the problem of estimating software and hardware performance at a high level, necessary to enable design space exploration, while maintaining an acceptable level of accuracy. The proposed methodology is general enough to be applicable to several formalisms and co-design environments. It is based on a uniform modeling of the system components, where the performance of both hardware and software are expressed in terms of *CPI*, and specific techniques to estimate such values starting from high-level specifications have been discussed. To validate the methodology, a number of experiments have been carried out (the chapter has shown in detail two examples belonging to the validation set) considering four commercial microprocessors: the achieved accuracy of the estimates is typically above 90%. The estimations have been performed on a SUN *UltraSparc2* running at 256 MHz and equipped with 128 Mbytes *RAM*, and the static estimation time has been approximately of 1*KLOC/second*.

7

System-Level Partitioning

In the system-level co-design flow proposed in this book (Figure 7.1), the system design exploration step is divided into two iterative tasks: *partitioning and architecture selection*, and *timing co-simulation*. All the data produced in the previous steps of the flow are used to guide the process, together with additional information provided by the designer. Such information expresses the eventual *architectural constraints* (e.g., max number of GPP, max number of DSP, gates limitation for FPGA, etc.), the *scheduling directives* (e.g., processes priority), and the parameters of the *communication model* (e.g., the number of concurrent communications allowed).

This chapter describes in detail the partitioning methodology. Such a step explores the design space to identify feasible solutions, supporting also the selection of a heterogeneous multi-processor architecture (type and number of components that should be included) taking into account several issues (degree of affinity, communication cost, processing elements load, concurrency, physical cost, etc.). The output of this methodology is the allocation of the behavioral components on to the selected architectural components. Architecture selection and partitioning are influenced by performance requirements, implementation cost, and application-specific issues.

The timing co-simulation methodology (Chapter 8) considers the proposed heterogeneous multi-processor architecture and a high-level model for the communication media to model the system behavior through the behavior of the hardware and software parts. It evaluates the performance of the system and verifies its timing correctness.

This chapter is structured as follows: Section 7.1 introduces the general issues related to system-level partitioning and overviews the main approaches presented in the literature. Section 7.2 details the characteristics of the proposed partitioning approach (i.e., model, methodology, tool, and validation), while Section 7.3 presents some final considerations.

123

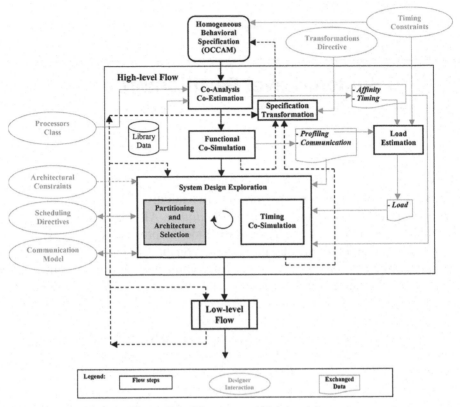

Figure 7.1   The proposed high-level flow.

7.1 Characterization

When the functionalities of the system have been proven correct by means of functional simulation or validation, they should be targeted for a specific implementation: hardware digital circuits or software programs. This task is called system-level partitioning.

In early design practice, partitioning was performed manually by experienced designers or system architects. In some cases, human intervention was supported by a set of cost figure estimates derived from static metrics or from simulation.

In the past years, researchers have primarily focused their interest on hardware-software co-design of *1-CPU-1-ASIC* architectures by proposing heuristics to move operations from hardware to software or vice versa to minimize costs and meet deadlines [127, 128].

The most critical problems in partitioning methodologies are related to the computational complexity, which is exponential with the number of modules (the partitioning problem is NP-hard), and to the difficulties of obtaining accurate estimates of the timing, silicon, or memory requirements and power consumption of the components.

Moreover, partitioning of multi-processors systems must solve not only hardware-software issues within a single embedded unit, but also the choice of the different parts composing the system (which may consist of hardware, software, or both) and the communication issues among them.

The design space that must be explored when considering all the parameters associated with a multi-processor architecture may become easily unmanageable, thus making impossible an exhaustive analysis of all the possible solutions. Even if only some of the main aspects are considered, such as system functionalities, number of processing elements, and hardware/software decomposition of the system or sub-systems that compose it, efficient heuristics are needed to manage the task complexity. The choice of the heuristics will probably conduce to a sub-optimal solution but this is the trade-off to be accepted to reach at least one feasible solution.

To solve the problem, typical multi-processor-oriented approaches [55, 129–133] are based on the following steps:

- Metrics definition;
- Cost function definition;
- Pre-allocation; and
- Optimization.

The first step aims at evaluating several indexes to point out specific features of the specification to be controlled or optimized during the partitioning. Such indexes are evaluated by means of estimation or analysis models. They provide a method to evaluate the system quality in relation with some specific parameters (e.g., area, performance, communication cost, etc.).

A proper combination of the defined metrics allows the definition of a cost function whose role is to provide a complete evaluation of the proposed solution quality, guiding the design space exploration toward solutions that minimize the cost function value, i.e., solutions that offer the best tradeoff between the considered aspects.

Most heuristic algorithms (e.g., *taboo search, simulated annealing, genetic algorithm*, etc.), in order to perform efficiently the design space exploration, require an initial valid solution to start from. The main goal is to find, in an efficient manner, feasible solutions that satisfy the imposed constraints by avoiding local minima of the cost function.

Finally, another important issue that influences all the steps of the partitioning task is the granularity level adopted to process the specification. Different partitioning strategies adopt different granularities: coarse-grain or fine-grain. The granularity indicates the basic element to be considered for allocation purposes. With a fine granularity, this element is typically the single statement of the specification; with a coarse one, it is typically a group of statements (e.g., a whole procedure). Finer granularities lead to more accurate results (increasing however the complexity of the partitioning itself). With a coarse granularity, it is possible to work to a higher abstraction level, thus obtaining a higher efficiency in the partitioning process. However, such increased efficiency is paid in terms of quality of the results; in fact, the results will heavily depend on the accuracy of the estimation and analysis models used to characterize the specification. In the co-design of heterogeneous multiprocessor embedded systems, a coarse granularity is the only chance to manage the system complexity.

The next section, based on the general characterization of the partitioning problem provided above, presents the proposed approach.

7.2 The Proposed Approach

The partitioning problem is defined as a mapping problem of a behavioral description of the system on a set of physical components. The algorithm allocates functionalities to processing elements while optimizing a global cost function. The algorithm is constituted by two phases: in the first phase, a clustering of the system functionalities is performed, based on qualitative and quantitative considerations, and a pre-allocation is obtained.

The second phase is constituted by refinements of such a pre-allocation, driven by a genetic algorithm and a global cost function that contains all relevant parameters to be taken into account. The cost function defined includes four main components: affinity (defined in Chapter 5), communication (as defined in Chapter 8), load (as defined in Chapter 8), and economical issues (next in this chapter). Thus, a variable-granularity design space exploration is performed to propose an architecture, and an allocation of the system functionalities on such an architecture, to meet all the specified constraints in an efficient way.

Existing tools, that handle multi-processor embedded systems co-design, start from a heterogeneous specification where the partitioning is practically suggested by the specification itself [129]. Others heavily rely on the experience of the designer [130, 131] or provide a partitioning methodology targeted only to particular applications [132, 133]. Some approaches focus mainly on communication issues [134, 135], while a general approach that considers both communications and different executors is presented in [136], but the *area-delay curves* considered in such a work are very difficult to use.

The approaches more similar to the one presented in this chapter are in [137, 138]. They perform a clustering of the behavioral specification, a task classification based on area and time, and a design space exploration by means of transformations. However, they are targeted to single processor systems.

7.2.1 Model and Methodology

This section details the methodology adopted to provide a solution to the partitioning problem, describing also the metrics, the cost function, and the algorithm used to explore the design space.

Problem Definition

Given the *OCCAM* system specification, the partitioning problem is related to the mapping of each functionality of the system (i.e., each OCCAM procedure) onto a heterogeneous multi-processor architecture. In particular, the output of this task is a description of the target architecture in terms of number and type of processing elements and the allocation of each system functionality on the considered processing elements. The solution provided should minimize communication and physical costs, while balancing the load on the processing elements. Furthermore, each functionality should be associated with the most suitable processing element type, in order to meet all the timing constraints defined by the user.

The optimal tradeoff is reached by means of a cost function that takes into account several metrics, evaluated by the analysis tools of the proposed co-design flow (Figure 7.1). In the following, such metrics and the adopted methodology are described in detail.

Model

During the design space exploration, the value of the cost function indicates the quality of the considered solution with respect to several parameters. In the following, such parameters, the related metrics, and the cost function are analyzed in detail.

Affinity Index

Given a solution, it is very important to evaluate how much it exploits the matching between the properties of the functionalities and the processing elements on which they have been allocated.

Therefore, based on the metric defined in Chapter 5, it is possible to evaluate the *Affinity Index* (I_A) of a solution as follows:

$$I_A = \frac{\sum\limits_{m \in \mathrm{MI}} [x_{\mathrm{GPP}m} \cdot (1 - A_{\mathrm{GPP}m}) + x_{\mathrm{DSP}m} \cdot (1 - A_{\mathrm{DSP}m}) + x_{\mathrm{HW}m} \cdot (1 - A_{\mathrm{HW}m})]}{|\mathrm{MI}|}$$

where:

- $x_{\mathrm{GPP}m}$, $x_{\mathrm{DSP}m}$, $x_{\mathrm{HW}m}$ is 1 or 0, respectively, if the method instance $m \in \mathrm{MI}$ is allocated or not to the associated type of executor;
- $A_{\mathrm{GPP}m}$, $A_{\mathrm{DSP}m}$, $A_{\mathrm{HW}m}$ are the affinities of $m \in \mathrm{MI}$ (Definition 15, Section 5.2.1);
- $|\mathrm{MI}|$ is the cardinality of MI (Definition 11, Section 4.3.2).

The *Affinity Index* values belong to the interval [0, 1]. Values closer to 0 indicate a perfect match between functionalities and executors.

Load Indexes

For each solution, it is very important to determine whether the exploitation of the proposed architecture is maximized by considering also load issues. This means that the load on the architecture should be balanced as much as possible and that no processing elements, if possible, should be overloaded or under-loaded.

To such purpose, two indexes are considered, to differentiate the software case from the hardware case.

The *Load Index* (I_{Lsw}) for software executors is evaluated as follows:

$$I_{Lsw} = \sum_{j=1..m} \frac{\frac{\sum_{i=1..nj} |li,j - L_{SW}|}{n_j \cdot L_{SW}}}{m}$$

where:

- m is the number of software executors present in the solution (GPP or DSP);
- n_j is the number of method instances of *MI* allocated on the software executor j;
- $l_{i,j}$ is the estimated load (Section 8.2.2) of the i-th functionality on the software executor j; and
- L_{SW} is the ideal load. It is theoretically 100% but it is generally set to approximately 70% (a typical value that allows considering the presence of a possible operating system [141]).

Such index values belong to the interval [0, 1]. It evaluates the average of the differences between the actual load and the real one. Values closer to 0 indicate that each processing element presents a load similar to the ideal one.

The load index for the hardware executors allows avoiding overload of hardware devices, i.e., the allocation of a number of functionality whose implementation will exceed the number of allowed gates [97]. Such an index (I_{Lhw}) is defined as follows:

$$I_{\mathrm{Lhw}} = \sum_{j=1..m} \frac{\frac{\sum_{i=1..nj} s_{i,j}}{n_j \cdot G_j}}{m}$$

where:

- $s_{i,j} = \begin{cases} 0 & g_{i,j} < G_j \\ g_{i,j} - G_j & g_{i,j} > G_j \end{cases}$
- m is the number of hardware devices (*ASIC* or *FPGA*);
- n_j is the number of method instances of *MI* allocated on the hardware executor j;
- $g_{i,j}$ is the number of gates needed to implement the i-th functionality on the executor j; and
- G_j is the maximum number of gates for the hardware executor j.

The load index values for hardware executors belong to the interval [0, 1] and it is 0 if no hardware executors are overloaded.

Communication Index

The information on communication cost (i.e., *procedure calls* and *channel communications*), gathered during the functional co-simulation (Section 8.2.2), allows the evaluation of the exchanged data size due to the interactions between functionalities allocated on different processing elements.

For this purpose, the *Communication Index* (I_C) is defined as follows:

$$Ic = \frac{\sum_{i \in \mathrm{MI}} \sum_{j \in \mathrm{MI}} com_{i,j}}{\sum_{i \in \mathrm{MI}} \sum_{j \in \mathrm{MI}} b_{i,j} \cdot c_{i,j}} \quad \text{with } i \neq j$$

where:

- $com_{i,j} = \begin{cases} 0 & E_i = E_j \\ b_{i,j} \cdot c_{i,j} & E_i \neq E_j \end{cases}$
- $|MI|$ is the cardinality of MI (Definition 11, Section 4.3.2);
- $b_{i,j}$ is the size of the data exchanged between $I \in MI$ and $j \in MI$;

- $c_{i,j}$ is the number of interactions between $i \in$ MI and $j \in$ MI; and
- E_i and E_j are, respectively, the executors of $i \in$ MI and $j \in$ MI.

The communication index values belong to the interval $[0, 1]$. Values closer to 0 indicate that the communication between functionalities allocated on different executors are negligible with respect to the total number of communications in the system.

Physical Cost Index

Another important parameter to be considered when evaluating a possible solution to the partitioning problem is the physical cost of the components present in the proposed architecture.

The *Physical Cost Index* ($I_\$$) is evaluated as follows:

$$I_\$ = \frac{\sum_{j \in E} \$j}{|\text{MI}| \cdot \$_{\text{MAX}}}$$

where:

- $|\text{MI}|$ is the cardinality of MI (Definition 11, Section 4.3.2);
- E is the set of the executors present in the proposed architecture;
- $\$_j$ is the physical cost of the executor $j \in$ E; and
- $\$_{\text{MAX}}$ is the cost of the most expensive executor $j \in$ E;

The *Physical Cost Index* values belong to the interval $[0, 1]$ and represent the cost of the proposed solution with respect to the most expensive one. Values closer to 0 indicates a cheaper solution.

Cost Function

By combining the metrics described above, it is possible to build a cost function that includes into all the aspects detailed and allows, during the design space exploration, a comparison between different solutions identifying the one that better represents a tradeoff between the different parameters. In fact, the affinity and load parameters tend to separate the functionalities in order to balance the load and exploit the processing elements features, while communications and physical cost tend to keep together such functionalities in order to minimize the number of processing elements.

Such considerations can be taken into account through a linear combination of the indexes, thus obtaining the following cost function expression:

$$\text{CF} = w_\text{A} \cdot I_\text{A} + w_\text{Lsw} \cdot I_\text{Lsw} + w_\text{Lhw} \cdot I_\text{Lhw} + w_\text{c} \cdot I_\text{C} + w_\text{s} \cdot I_\text{S}$$

where $w_A, w_{Lsw}, w_{Lhw}, w_c,$ and $w_\$$ are the weights associated with the different parameters, thus providing the possibility of assigning a different importance to the different parameters.

Methodology

The proposed methodology explores the design space by using a genetic algorithm. This algorithm allocates functionalities to processing elements while minimizing the cost function presented above. The methodology works with a variable coarse-grain granularity where a functionality or a group of functionalities is allocated to the same processing element.

The methodology is composed of several steps that are iteratively repeated, working each time with a finer granularity, until a solution to the problem is found. Figure 7.2 shows the steps of the algorithm, detailed in the following:

Clustering

The clustering represents the starting point of the methodology. Its purpose is to select groups of functionalities, bounded by the caller-callee relationship, to be considered as clusters, i.e., not to be further decomposed, in the following steps of the same iteration of the methodology.

In this way, each iteration works with a finer granularity. This reduces the complexity of the analysis, since the first iterations involve a limited number of clusters. This number increases if a feasible solution is not found, thus

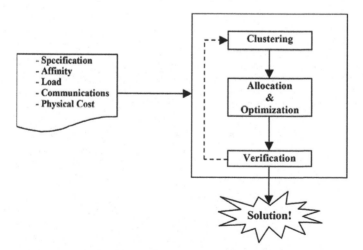

Figure 7.2 Partitioning methodology steps.

requiring a more detailed analysis of the elements composing the clusters. Since each cluster is allocated on the same processing element, the former solutions tend to be the cheaper, while the latter tend to better exploit the characterization of each single functionality with respect to the type and number of executors.

As an example of clustering, let us consider the *Procedural Interaction Graph (PING*, Section 4.3.2) presented in Figure 7.3.

The first iteration of the methodology proposes only a cluster that groups all the instances of the method. The second one considers the nodes identified by the *non-blocking calls* (i.e., F1, F6, F9; this implicitly considers the exploitable concurrency, Section 4.3.2) and creates four clusters: {Main}, {F1, F2, F3, F4, F5}, {F6, F7, F8}, and {F9, F10, F11}. The following iterations follow a *depth-first* decomposition (that can be guided by the designer or by the timing constraints) and proceed to consider each possible combination limited by the *all-or-1* rule: i.e., each instance of a method is in a cluster with all the callee instances or with none one of them. For example, considering the tree starting from F1, the clusterings are only three: {F1, F2, F3, F4, F5}, {F1}{F3}{F2, F4, F5}, and {F1}{F2}{F3}{F4}{F5}. In this way, the complexity of the methodology is limited, preserving however the meaningfulness of the provided clustering. Finally, the last iteration works considering each instance of a method as a single cluster.

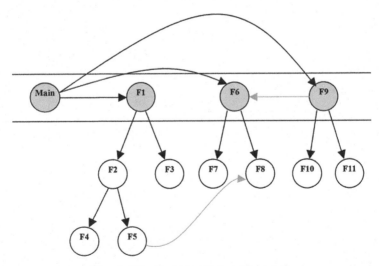

Figure 7.3 Procedural interaction graph.

Allocation and Optimization

This step allocates the clusters provided by the previous steps to different processing elements by exploring the design space in search of the allocation that minimizes the cost function previously defined. This step is based on a genetic algorithm [139], where each individual of the considered population represents a possible architecture/allocation item. The structure of the individual is represented by an entry for each cluster considered during the current iteration (Figure 7.4): each cluster is associated with a type of processing element and an instance of it (the maximum number of instances allowed for each type of processing element is specified by the designer).

The initial population is randomly generated, while, during the evolution of the population, the algorithm performs the optimizations that minimize the cost function following the classical rules of genetic algorithms [139]. The *crossover* operation generates two new individuals combining two existing ones as shown in Figure 7.5 while the *mutation* operation changes randomly type and instance number of the processing element associated with a randomly selected cluster.

During the evolution, the individuals that score the worst values tend to be replaced by better ones. Several parameters in the algorithm (e.g.,

Cluster 1	Cluster 2	Cluster 3	...	...	Cluster n
GPP	HW	GPP	GPP	DSP	GPP
1	1	3	2	1	1

Figure 7.4 Individual structure.

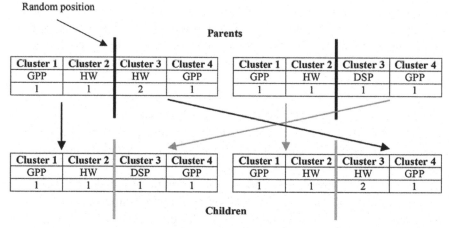

Figure 7.5 Crossover.

population size, number of generations, mutation probability, etc.) allow a wide exploration of the solution space, while avoiding local minima.

Verification

The architecture/allocation solution provided by the previous step is verified by means of the *Timing Simulation* (such a simulation contributes also to define the interconnection network and the scheduling policy, as fully described in Chapter 8). If such a combination is not a valid solution to the partitioning problem (i.e., it does not meet the timing constraints), the whole process is repeated starting from a finer-granularity clustering. In this way, the proposed solution makes use of more resources (i.e., it is more costly), but it allows considering with more detail the characteristics of the specification better exploiting the feature of the executors, producing a target architecture more tailored to the problem, with an higher probability to satisfy the constraints.

7.2.2 The Tool

The proposed partitioning tool, called E_{MuP} (*Embedded Multi-processor Partitioning*), has been developed following two different goals: to integrate the tool in the *TOHSCA* environment and to provide an easy portability toward different co-design environment.

The tool has been developed in C++ and it is based on a library of classes for genetic algorithms (*GALIB* [140]). The *portability-enabler* feature is the input format accepted by the tool: it is based on the *VCG* format [84], a third-party format that can be managed and visualized with third-party open source tools.

Starting from an annotated (with affinity, load, and communication data) VCG; E_{MuP} builds its procedure-level internal model (Section 4.3.2). Both the VCG and the internal model are suitable to represent specifications expressed in several specification languages enabling the tool to be integrated in different co-design environments. In the case of TO(H)SCA, the VCG file is generated by the co-simulator (i.e., during the functional simulation) and annotated with the data generated during the previous steps.

Therefore, the internal model allows the tool to perform the cost function minimization based on the rule of genetic algorithms as described in the previous paragraph.

The output of the partitioning tool is both a simple association list (i.e., a text file) between each instance of method and a processing element type and instance, and a VCG graph representing the proposed architecture and allocation. The list can be easily considered by a co-simulator engine in order to verify the quality of the proposed solution while the VCG can be visualized and thus it is useful for the designer intervention.

7.2.3 Validation

This paragraph shows the effectiveness of the proposed system-level partitioning methodology when a system design exploration has to be performed.

The example considered is composed of 52 *OCCAM* procedures and its *Procedure Interaction Graph* (Section 4.3.2) is represented in Figure 7.6. It is not a real application but it has been obtained by composing (and coding with the *OCCAM* language) some of the procedures belonging to the test suite described in Section 5.2.3. In this way, it has been possible to create an example with a homogeneous distribution of the affinity values. The target architecture is composed of an unconstrained number of *GPP*, *DSP*, and *FPGA*.

Table 7.1 shows the affinity values (Chapter 5) of each procedure. The load depends on the imposed timing constraints and the relative physical costs are 1 for a GPP, 1.2 for a DSP, and 2 for a FPGA.

The goal of this validation is to check the behavior of the partitioning tool for different timing constraints and different cost function weights in order to

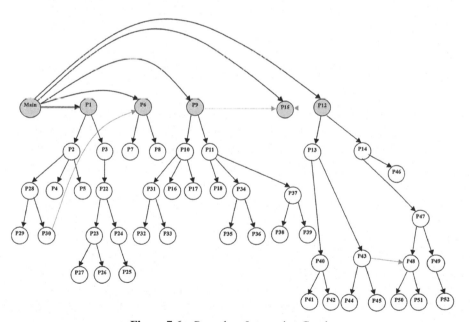

Figure 7.6 Procedure Interaction Graph.

Table 7.1   Affinity values

Procedure	Affinity			Procedure	Affinity		
	GPP	DSP	HW		GPP	DSP	HW
P1	0.612	0.605	0.397	P27	0.799	0.447	0.492
P2	0.425	0.649	0.388	P28	0.580	0840	0.403
P3	0.535	0.640	0.392	P29	0.460	0.840	0.403
P4	0.259	0.772	0.398	P30	0.559	0.772	0.398
P5	0.297	0.748	0.396	P31	0.868	0.046	0.688
P6	0.543	0.551	0.468	P32	0.597	0.053	0.911
P7	0.658	0.062	0.894	P33	0.997	0.053	0.211
P8	0.997	0.053	0.925	P34	0.643	0.571	0.416
P9	0.586	0.619	0.394	P35	0.580	0.620	0.405
P10	0.548	0.640	0.391	P36	0.544	0.653	0.396
P11	0.351	0.637	0.393	P37	0.608	0.595	0.406
P12	0.623	0.577	0.413	P38	0.520	0.659	0.388
P13	0.553	0.626	0.397	P39	0.553	0.634	0.390
P14	0.424	0.648	0.388	P40	0.604	0.911	0.409
P15	0.524	0.648	0.388	P41	0.471	0.810	0.401
P16	0.997	0.053	0.915	P42	0.352	0.748	0.396
P17	0.612	0.596	0.412	P43	0.459	0.772	0.388
P18	0.587	0.607	0.405	P44	0.259	0.772	0.398
P19	0.616	0.609	0.397	P45	0.574	0.651	0.533
P20	0.725	0.519	0.428	P46	0.384	0.648	0.388
P21	0.565	0.635	0.398	P47	0.524	0.628	0.368
P22	0.997	0.053	0.550	P48	0.234	0.608	0.378
P23	0.658	0.053	0.694	P49	0.987	0.063	0.921
P24	0.799	0.447	0.492	P50	0.992	0.092	0.890
P25	0.494	0.648	0.388	P51	0.968	0.085	0.787
P26	0.384	0.648	0.388	P52	0.989	0.099	0.833

highlight, in particular, the role of the affinity values and the load balancing issues, while minimizing the physical and communication costs of the system.

The ideal load L_{SW} considered in the load index has been set to 70% (a typical value [141] that allows to consider the presence of a possible operating system). Different timing constraints have been imposed on the execution time of the whole application in the following way: with respect to a T_{REF} (evaluated by simulation for a single *GPP* system) in different experiments, the constraints have been 90% T_{REF}, and 50% T_{REF}. The latter constraint aims at forcing the partitioning tool to exploit the concurrency by increasing the number of executors.

The results described in the following have been obtained with the weights shown in Table 7.2. The values of the weights have been set with the aim of increasing the importance of the load balancing issues with respect

Table 7.2 Cost function weights

Index	Weight
I_A	Variable
I_{Lsw}	4.0
I_{Lhw}	4.0
I_C	2.0
$I_\$$	2.0

to communications and physical costs: this forces the partitioning tool to carefully consider the use of the resources. The weight of the affinity is variable and it assumes different values during several experiments. In detail, these values are $\{0, 2, 3, 4, 7\}$ allowing to enforce at each step weights of the affinity index in the cost function.

For each value of the affinity weight, Tables 7.3 and 7.4 report the iteration (each iteration starts with a finer-granularity clustering) that has found the minimum value for the considered cost function and the related timing simulation result.

Table 7.3 shows the results for the constraint 90% T_{REF}. The constraint forces an architecture with more than one executor in order to reduce the execution time and, in fact, the timing constraint is always largely met. With lower affinity weights (e.g., 0 and 2), the partitioning does not consider *DSPs* executors, while with the weights 3 and 4, the affinity becomes an important factor and a DSP is introduced. Theses solutions provide good simulated times

Table 7.3 Timing constraint: 90% T_{REF}

w_A	Iteration	I_C	I_{LSW}	I_A	Architecture GPP	DSP	FPGA	Simulated Time
0	6	0.003	0.004	0.501	1	0	1	66% T_{REF}
2	6	0.003	0.121	0.395	2	0	0	69% T_{REF}
3	9	0.020	0.121	0.390	1	1	0	60% T_{REF}
4	5	0.004	0.126	0.384	1	1	0	58% T_{REF}
7	9	0.032	0.277	0.254	2	1	0	55% T_{REF}

Table 7.4 Timing constraint: 50% T_{REF}

w_A	Iteration	I_C	I_{LSW}	I_A	Architecture GPP	DSP	FPGA	Simulated Time
0	10	0.320	0.040	0.394	3	0	0	51% T_{REF}
2	9	0.006	0.024	0.386	2	1	0	42% T_{REF}
3	9	0.007	0.024	0.385	2	1	0	42% T_{REF}
4	9	0.011	0.126	0.392	2	1	0	43% T_{REF}
7	10	0.220	0.060	0.340	1	2	0	45% T_{REF}

and an acceptable physical cost. When the affinity weight is too high, the tool considers the affinity more than other factors and then, even if the simulated time is the best one, the physical cost increases and the load index indicates an unbalancing that indicates a possible under-load of the resources. In this example, the timing constraint is not considered as a big issue.

Table 7.4 shows the results for the constraint 50% T_{REF}. The heavy constraint forces an architecture with several executors. The timing constraint is always met except in the case when the affinity index is not taken into account (i.e., $w_A = 0$). With affinity weights from 2 to 4, the partitioning provides good solutions. Finally, as in the previous case, an affinity weight too high arrives to solutions that do not consider properly the other aspects: in this case, communication issues cause a worst simulated time.

The results show as the partitioning tool is able to perform effective design space exploration, while the affinity represents, when considered important as other aspects, a useful indicator that allows the selection of an architecture tailored to the features of the specification. Finally, the examples have shown, as an integrated co-simulator engine (Chapter 8) allows performing, an effective system design exploration step.

7.3 Conclusion

This chapter has shown the *partitioning and architecture selection* task that, with the *timing co-simulation* one, compose the system design exploration step of the proposed co-design flow.

More in detail, after introducing the main partitioning issues, the metrics and the cost function adopted in the proposed approach have been accurately defined showing the interaction with the other tools of the environment. Then, the methodology, based on an initial *clustering* and on a heuristic optimization step, has been analyzed in detail.

Two final examples have shown the effectiveness of the proposed approach, enforcing the meaningfulness of the concept of affinity (Chapter 5) and showing the possible interactions between the partitioning tool and the co-simulator engine (Chapter 8). To perform such system design explorations, the partitioning tool, implemented in C++ and executed on a *Pentium III* running at 700 MHz equipped with 640 Mbytes *RAM,* has processed a population of 500 individual for 2500 generations in a time shorter than 30 min (comprehensive of the co-simulation performed on a *SUN UltraSparc2* running at 256 MHz and equipped with 128 Mbyte).

8

System-Level Co-Simulation

Aim of this chapter is to present the related co-simulation step and its role in the global design flow. In fact, co-simulation is involved in several steps of the proposed flow (Figure 8.1) in order to perform dynamic analysis on the system features, check the functional correctness of the specification, and check the quality of the solution provided by the partitioning and architecture selection tool. In particular, two types of co-simulation are considered: *functional* and *timing* one.

The *functional simulation* allows checking the system functionalities to verify their correctness with respect to typical input data sets. This kind of simulation is very fast allowing the designer to easily detect functional errors. In fact, the timing aspects are not considered but other issues, like synchronization, precedence, and mutual exclusion between *OCCAM* processes can be observed, possibly detecting anomalous situations in such a deadlock or the presence of dead code. Moreover, after static analysis and estimation, it is possible to extract other important data characterizing the dynamic behavior of the system: *profiling* and *communication cost.*

This means that it is possible to evaluate the number of executions of each *OCCAM* process, the amount of data exchanged between procedures, and the set of procedures that typically run concurrently in the system (these information are always related to the behavior of the system in correspondence to typical input data sets). Finally, the early detection of anomalous behavior allows the designer to correct the specification, thus avoiding a late discovery of problems that could lead to time-consuming (i.e., costly) design loops.

Combining some of the data provided by the previous steps of the design flow (static timing estimations and profiling) with the designer imposed timing constraints allows the estimation of the *load* that each OCCAM procedure will impose to a processor (*GPP*) that should execute it. The extraction of these data from a behavioral specification is an important task that allows, during the system design exploration step, the evaluation of the number of

139

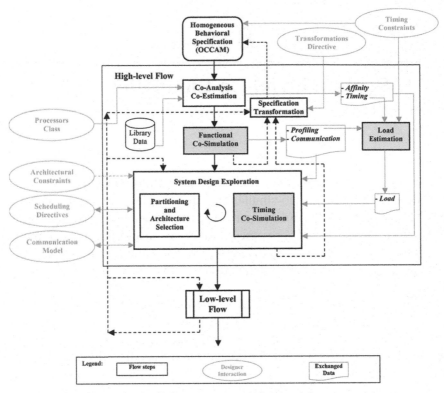

Figure 8.1   The proposed high-level flow.

needed processing elements and the identification of those procedures that will probably need an executor more performing than a GPP.

The *timing co-simulation*, together with the *partitioning and architecture selection*, is part of the system design exploration step. All the data produced in the previous steps (co-analysis, co-estimation, and functional simulation) are used to guide the task, together with additional information provided by the designer. The timing co-simulation methodology considers the proposed heterogeneous multi-processor architecture and a high-level model for the communication media in order to model the system behavior through the behavior of the hardware and software parts. It evaluates the performance of the system by verifying its timing correctness. Moreover, it allows modification in the communication model parameters and the designer intervention to set different scheduling directives effectively, thus supporting the design space exploration.

This chapter addresses the system-level co-simulation step and it is structured in the following way. Section 8.1 describes the general co-simulation issues, Section 8.2 describes the proposed approach, by analyzing in detail the model and the methodology that characterize it, and Section 8.3 shows the result obtained through the application of the proposed approach.

8.1 Characterization

The need to specify and simulate mixed hardware/software embedded systems possibly at a high abstraction levels has been recognized in many application fields, such as multimedia [5], automotive [6], mechatronics [7], and telecommunication [8, 9]. A key point in such an activity is the possibility of taking most of the decisions during the earlier stages of the design, to avoid as much as possible design loops including time-consuming synthesis activities. A high-level co-simulation is the ideal platform where the designer validates the system functionality and different tradeoff alternatives.

The specification of a hardware/software system usually follows one of the approaches described below. Both suffer some limitations related to system simulation and verification.

- **Single-language**: the whole system is described using a single language. This approach simplifies specification and verification, but poses severe limitations on the simulation efficiency. This is mostly because no languages, currently, are available to efficiently model strongly different and heterogeneous modules.
- **Multi-language**: modules are modeled using the most appropriate language. This leads to an effective simulation of the stand-alone components but requires an environment [66] capable of efficiently implementing the communication among the different simulation engines.

One major limitation of the single-language approach depends on the difficulty of modeling with acceptable accuracy the hardware portion of the design and the microprocessor running the software. A *HDL* model of the microprocessor is hardly acceptable: at the behavioral level, it does not provide cycle-accurate results that are often necessary; at the *RT*, or gate-level, when available, it requires long simulation times.

When the multi-language strategy is adopted, the hardware part of the system, described in one of the many *HDL*, is simulated with commercial tools while the microprocessor running the software program is modeled using *ad-hoc* languages such as *MIMOLA* [67, 68], and *nML* [66, 69, 70] that provide the required efficiency and accuracy.

The simulation of a complex system has a two fold purpose. On one hand, it aims at verifying whether the functionality provided by the model is compliant with the specification; on the other hand, it has the goal of producing measures of the system that are essential for the following phases of partitioning and synthesis. When accurate estimates of timing, area and/or power are available, simulation is also a valuable tool to perform a preliminary design space exploration aiming at evaluating different design alternatives.

A number of co-simulation strategies are emerging in the literature [5, 71, 72, 11, 73] and commercially (e.g., Coware [74], Seamless [75]), to cope with the problem of building a virtual prototype of the system, where the different descriptions of the blocks can be executed. At higher abstraction levels, the tools can achieve high simulation speed, but their purpose is mainly related to the functional verification of the system behavior and component interactions. To obtain a more detailed analysis on the system behavior (timing, communication, and concurrency), it is necessary to perform the analysis at lower level of abstraction back annotating the result at the higher levels. The proposed approach, described in the next chapter, thanks to the full integration with the other TOHSCA system-level tools allows a complete characterization of the system behavior, thus avoiding time-consuming design loops between abstraction levels.

8.2 The Proposed Approach

A crucial problem in designing multi-processor embedded systems is the possibility of efficiently comparing the timing behavior of different system configurations to select, during the early stages of the design process, a suitable tradeoff between performance and cost. The goal of this chapter is to present a timing system-level simulation strategy allowing the user to simulate, in a flexible and effective manner, a multi-processor embedded architecture enabling the user to evaluate the impact of modifying the main configuration parameters of the overall system. The relevant aspects a user can inspect to explore the design space concern the communication architecture, the process-scheduling policy and the type of the processing elements (*GPP, DSP,* and *ASIC/FPGA*).

The value added of the proposed simulation strategy is the possibility of validating the behavior of the overall system while exploring the design space, considering aspects concerning both hardware/software and software/software partitioning. The TOHSCA system-level co-simulation approach is the cornerstone to perform synthesis decisions. Such a task normally implies a high computational effort: each process needs to be compiled, producing object code or synthesized hardware before the evaluation and the back annotation of the results on the system description. Moreover, these evaluation steps have to be carried out each time the technology or the architecture is modified (e.g., hw/sw partition, instruction set, technological libraries, etc.) making the design space exploration costly. To overcome these drawbacks, it is beneficial to estimate the performance by operating at the highest possible abstraction level. For this purpose, a methodology has been developed to implement the proposed hw/sw analysis models within the TOHSCA co-simulation engine. The developed tool makes able the designer to perform fast *functional simulations* to efficiently check the functional correctness of the specification, while *timing simulations* are possible, thanks to the integration with the estimation tool (Chapter 6), to verify the satisfaction of the timing constraints, defined by the user.

8.2.1 Model and Methodology

This Paragraph introduces the problem of heterogeneous multi-processor embedded systems timing co-simulation, recalls the timing model adopted in the proposed co-simulation methodology, and presents the methodology itself, focusing on the main algorithms that constitute the kernel of the simulator engine.

Problem definition

The proposed methodology aims at providing multi-processor embedded systems co-design capability to an existing single-processor co-design environment (i.e., TOSCA, Section 2.1) with particular emphasis to the co-simulation task at the specification-level. In fact, the relative weight of the co-simulation performed during the early stages of the design flow is considered particularly relevant since it allows the designer a fast evaluation of the timing properties of the system before moving down to the detailed analysis/synthesis of the selected solution. The benefit of this approach is twofold: first, it is possible to discard the architectural alternatives not satisfying the cost-performance requirements during the early stages of the design flow, and second, it is possible to efficiently support the partitioning task, while remaining at a high level of abstraction.

The system specification, which is given by using the OCCAM language extended to include timing constraints, is composed of a set of simple processes (an OCCAM *process* corresponds to a single statement, e.g., an assignment) grouped into procedures. The communication between concurrent processes is explicit and it is based on the *rendezvous* model (*OCCAM channels*). An important advantage of using channels to represent the communication is the possibility of applying the same compact model of data exchange between heterogeneous processes (and consequently interfaces).

Hardware/Software timing model

The proposed timing simulation environment takes into account three factors:

- the execution time of each statements;
- the impact of using more than one processing element; and
- the effects of communication.

While communication effects and microprocessor load are evaluated dynamically during the simulation by using, as a reference, an appropriate set of models presented in the following, the execution time is statically estimated off-line, as described in Chapter 6, and dynamically used when the simulation is carried out.

System-level co-simulation methodology

The proposed approach performs the system-level co-simulation of heterogeneous embedded systems composed of several processing elements working concurrently where the processing elements can be a mix of microprocessors (general purpose or specialized) and co-processors (i.e., customized HW components implemented as ASIC or FPGA). The presented co-simulation methodology is mainly characterized by two entities: an *execution model* and a *system-level communication model*.

The first one, whose implementation is identified in the *time-stretching algorithm*, models both the *physical parallelism*, due to the presence of several processing elements, and the *virtual parallelism*, due to the presence of concurrent processes running on the same computational resource; it is worth noting that the execution model is parametric with respect to the scheduling policy.

The second one provides a high-level model for the communication architecture that takes into consideration the overhead due to both the data transfer and the use of the shared physical resources (the cooperation or the interference among processes implies an inevitable conflict during the access to the shared resources). Both these aspects are analyzed in depth in the following.

Execution model

The *execution model*, which represents the kernel implemented in the event-driven simulator, gathers the architectural aspects, in terms of processing elements, and the functional aspects, in terms of processes and timing constraints. To take into account the effects associated with the *virtual parallelism*, the *execution model* extends (*stretch*) the execution time of each process included in the ready queue according to both the CPU workload and the scheduling policy.

The contribution of this book is related to the extension of the algorithm reported in [11] to multi-processor systems; in particular, the extended version of the algorithm is parametric with respect to both the number of processors and the scheduling policy.

Notations

The following notations (Figure 8.2) support the presentation of the algorithm corresponding to the execution model.

i:	Process index
j:	Execution element index
T_{abs}:	Time instant corresponding to the actual behaviour of the system under simulation
$t_{intrinsic}(i)$:	Execution time of process I by considering unlimited resources in case of hardware-bound and a dedicated CPU in case of software-bound (clock cycles).
$t_{remaining}(i)$:	Time to complete the execution of the process i in the case it is the only active process (clock cycles)
$t_{expanded}(i)$:	Time to complete the execution of the process i considering workload and scheduling policy (clock cycles)
$n_A(j)$:	Number of processes active on processor j
$n_{AH}(j)$:	Number of high priority processes active on processor j
$n_{AL}(j)$:	Number of low priority processes active on processor j
$\tau_c(j)$:	Context switching time on processor j
$T_c(j)$:	Commutation period on processor j
$n_I(j)$:	Number of idle processes on processor j
$T_I(j)$:	Spooling period on processor j
$OH(j)$:	Overhead on processor j due to the management of low-priority processes ($0<OH(j)<1$)
M:	Time interval between the events k and $k+1$
$d(j)$:	Dilatation coefficient on processor j if a round-robin scheduling policy is used for all the processes
$d_H(j)$:	Dilatation coefficient on processor j for processes with high priority
$\Delta t(j)$:	Overhead to be added to low-priority processes execution time on processor j

Figure 8.2   Notations.

Multi-processors *time-stretching* algorithm

The time evolves proceeding by discrete intervals of time M_k, where k is the k-th step of the algorithm. This simple model of time is regulated by events corresponding to the start or the end of either a process execution or inter-processes activities (e.g., loading of data on a channel). M_k is computed by analyzing the events queue. Events are sorted by evaluating the execution time of the active processes; more precisely, in order to emulate the effects of the processing element workload and of the scheduling policy, the execution time of the software processes are *expanded* (according to its priority and processing element). For example, it is reasonable to think that, if $n_A(j)$ processes are active on the j-th processor, each process is slowed down $n_A(j)$ times. Once the evolution of time has been determined, the new *remaining time* of each active software process has to be evaluated; this operation is computed by *contracting* the difference between the expanded old remaining time and M_k. In case of a hardware process, the *remaining time* is the difference between the previous *remaining time* and the interval M_k: no contraction is required. This procedure is iterated until a process is active. Figure 8.3 reports a schematic representation of the procedure applied to three software processes, with the same priority, running on the same processing element. Without considering priorities, the *execution model* is mainly characterized by the dilation (i.e., expansion) coefficient on processor j (a round-robin scheduling policy is used for all the processes on processor j):

$$d(j) = n_A(j) \cdot \left(1 + \frac{\tau_c(j)}{T_c(j)}\right) + n_I(j) \cdot \frac{\tau_I(j)}{T_I(j)}$$

where $\frac{\tau_c(j)}{T_c(j)}$ models the overhead due to the context switch of the active software processes and $\frac{\tau_I(j)}{T_I(j)}$ identifies the overhead, due to the operating system, to manage the idle communication processes.

Consequently, the relation between the *expanded time* and the *compressed time* of each i-th software process is determined by the equations:

$$t_{\text{expanded\_}k}(i) = d(j)_k \, t_{\text{remained\_}k}(i)$$
$$t_{\text{remained\_}k+1}(i) = t_{\text{remained\_}k}(i) - M_k$$

where k is the k-th step. It is worth noting that, when the process starts the remaining time coincides with the intrinsic time ($t_{\text{intrinsic}}(i)$).

For example, by considering a scheduling policy with two levels of priority, it is possible to evaluate the dilation coefficient for high-priority processes:

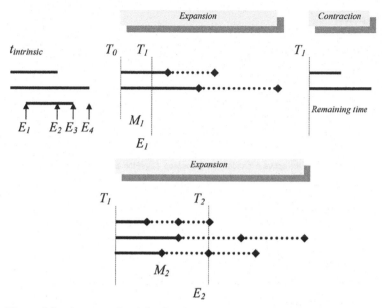

Figure 8.3   An example of the time-stretching procedure for three processes.

$$d_{\mathrm{H}}(j) = n_{\mathrm{AH}}(j) \cdot \left(1 + \frac{\tau_c(j)}{T_c(j)}\right) + n_I(j) \cdot \frac{\tau_I(j)}{T_I(j)} + \delta_{n_{\mathrm{AL}}(j)} \cdot \mathrm{OH}(j)$$

where OH(j) is the overhead on processor j to manage low-priority processes (it is the percentage of the CPU time dedicated to low priority processes) while $\delta_{n_{AL}(j)}$ equals 1 if the number of low priority processes is greater then zero (otherwise it returns 0). The overhead to be added to the low-priority processes execution time, evaluated considering the extra time dedicated to the high-level processes with respect to a round-robin approach, is

$$\Delta t(j) \quad = \quad \frac{(d(j) - d_{\mathrm{H}}(j)) \cdot \sum_{i=1\ldots nAH} t_{\mathrm{intrinsic}}(i)}{n_{\mathrm{AL}}} \cdot \delta_{n_{AL}(j)}$$

It is worth noting that the model, and therefore the algorithm, can be modified to consider more priority levels to the detriment of the computational complexity. Figure 8.4 shows the complete algorithm.

System-level communication model

The inter-processes communication occurs through blocking channels (*rendezvous*). By considering the nature of such a communication mechanism,

∀ active process *i*,
⇒ $t_{remaining}(i) = t_{intrinsic}(i)$;

while (*exists active processes*)
{
        ∀ processor *j*,
                ⇒ evaluate $d(j)$, $d_H(j)$ and $\Delta t(j)$;

        ∀ active high-priority process *i* on processor *j*,
                ⇒ $t_{expanded}(i)=t_{remaining}(i)^*d_H(j)$;

        ∀ active low-priority process *i* on processor *j*,
                ⇒ $t_{expanded}(i)=t_{remaining}(i)^*d(j)+\Delta t(j)$;

        M = min(t_{min} of all the active expanded SW processes, t_{min} of the active HW processes);

        $T_{abs} = T_{abs} + M$;

        ∀ active HW process *i*,
                ⇒ $t_{remaining}(i) = t_{remaining}(i) - M$;

        ∀ processor *j*,
        ∀ active process *i* on processor *j*,
                ⇒ $t_{remaining}(i) = (t_{expanded}(i) - M)/d(j)$;

        delete from the active processes pool the ones executed in correspondence of T_{abs} ;

        activate new processes ;

        ∀ new process *i* activated ⇒ $t_{remaining}(i) = t_{intrinsic}(i)$;
}

Figure 8.4   Multi-processors time-stretching algorithm.

different time components are used to model the communication procedure
and the contention for the communication resources. The *input* process has
a fixed intrinsic time $t_{intrinsic}$ related to the operations needed to prepare the
communication and a time t_{move} needed to move the data from the memory
to a buffer. Similarly, the *output* process is characterized by a fixed intrinsic
time $t_{intrinsic}$ related to the operations needed to complete the communication
and a time t_{move} required to move the data from the buffer to the memory. The
time necessary to move a single bit (t_{bit}) is assumed to be the same both for
all the input and output processes, and t_{move} is evaluated on the basis of the
size (s) of the data (expressed in bit) and the size (B) of the communication
resource [103, 104]:

$$t_{move} = \lceil s/B \rceil \cdot t_{bit}$$

These data are kept in the buffer while the communication lines are available
for the processes (t_{avail}). Finally, the transferring delay through the communication media is represented by the time t_{trans}: such a delay depends on the
data size and communication media features.

Table 8.1 Channel Communication times

	Input	Output
$sw_i - sw_i$	$t_{\text{intrinsic}} + t_{\text{move}}$	$t_{\text{intrinsic}} + t_{\text{move}}$
$sw_i - sw_j$	$t_{\text{intrinsic}} + t_{\text{move}} + t_{\text{avail}} + t_{\text{trans}}$	$t_{\text{intrinsic}} + t_{\text{move}} + t_{\text{avail}} + t_{\text{trans}}$
$sw - hw$	$t_{\text{intrinsic}} + t_{\text{move}} + t_{\text{avail}} + t_{\text{trans}}$	$(t_{\text{intrinsic}} + t_{\text{move}}) + t_{\text{avail}} + t_{\text{trans}}$
$hw - sw$	$(t_{\text{intrinsic}} + t_{\text{move}}) + t_{\text{avail}} + t_{\text{trans}}$	$t_{\text{intrinsic}} + t_{\text{move}} + t_{\text{avail}} + t_{\text{trans}}$
$hw - hw$	$(t_{\text{intrinsic}} + t_{\text{move}} + t_{\text{trans}})$	$(t_{\text{intrinsic}} + t_{\text{move}} + t_{\text{trans}})$
$hw_i - hw_j$	$(t_{\text{intrinsic}} + t_{\text{move}}) + t_{\text{avail}} + t_{\text{trans}}$	$(t_{\text{intrinsic}} + t_{\text{move}}) + t_{\text{avail}} + t_{\text{trans}}$

Since t_{avail} and t_{trans} are independent of the type of the processes involved in the communication (*hw-hw*, *sw-sw* and *hw-sw*), they are not influenced by the time-stretching procedure but these times are managed by the *Communication Manager*.

Table 8.1 summarizes the relation between the processes type, in terms of hardware or software (*i* and *j* indexes identifies the same or different processors), and the channel communication time. It is worth noting that, since hardware components are faster than software and considering that hardware modules communicate by means of dedicated lines, some time components (inside brackets in Table 8.1) are neglected. Such assumptions are typical and they have been proposed and validated for several heterogeneous system architectures [100–103].

Table 8.2 reports the same timing model for procedure calls. The main difference with respect to the channel communication is that the communication times are evaluated considering the size of both the *passed* (i.e., $t_{\text{move\_p}}$, $t_{\text{trans\_p}}$) and the *returned* ($t_{\text{move\_r}}$, $t_{\text{trans\_r}}$) parameters.

The time estimation for accessing the communication hardware resources (t_{avail}) and for transferring data (t_{trans}) is based on a parametric communication model; in particular, both times are computed based on the average number of concurrent communications allowed using the selected hardware resources (n_{ave}), and the average number of hops (h_{ave}) [45]. The advantage of this high-level abstraction model for the communications architecture is twofold: on one side, it allows the designer to ignore the final implementation details; on the other side, it allows a fast exploration of a broad range of architectures postponing the detailed definition of the communication network after the main system-level decisions have been frozen.

Table 8.2   Procedure call times

	Caller	Callee
$sw_i - sw_i$	$t_{\text{intrinsic}} + t_{\text{move\_p}}$	$t_{\text{intrinsic}} + t_{\text{move\_r}}$
$sw_i - sw_j$	$t_{\text{intrinsic}} + t_{\text{move\_p}} + t_{\text{avail}} + t_{\text{trans\_p}}$	$t_{\text{intrinsic}} + t_{\text{move\_r}} + t_{\text{avail}} + t_{\text{trans\_r}}$
$sw - hw$	$t_{\text{intrinsic}} + t_{\text{move\_p}} + t_{\text{avail}} + t_{\text{trans\_p}}$	$(t_{\text{intrinsic}} + t_{\text{move}}) + t_{\text{avail}} + t_{\text{trans}}$
$hw - sw$	$(t_{\text{intrinsic}} + t_{\text{move\_p}}) + t_{\text{avail}} + t_{\text{trans\_p}}$	$t_{\text{intrinsic}} + t_{\text{move\_r}} + t_{\text{avail}} + t_{\text{trans}}$
$hw - hw$	$(t_{\text{intrinsic}} + t_{\text{move}} + t_{\text{trans\_p}})$	$(t_{\text{intrinsic}} + t_{\text{move\_r}} + t_{\text{trans\_r}})$
$hw_i - hw_j$	$(t_{\text{intrinsic}} + t_{\text{move\_p}}) + t_{\text{avail}} + t_{\text{trans\_p}}$	$(t_{\text{intrinsic}} + t_{\text{move\_p}}) + t_{\text{avail}} + t_{\text{trans\_p}}$

As an example, let us consider three concurrent communications and an architecture where the available communication hardware resource is a single bus ($n_{\text{ave}} = 1$ and $h_{\text{ave}} = 0$). Since processes can be served one at a time, the first queued communication is served (its $t_{\text{avail}} = 0$) while the others have to wait.

By referring to this FIFO policy, the second communication has $t_{\text{avail}} = t_{\text{trans\_1}}$ while the third one has $t_{\text{avail}} = t_{\text{trans\_1}} + t_{\text{trans\_2}}$. Another example concerns $n_{\text{ave}} = 3$ and $h_{\text{ave}} = 0$ that models either a crossbar or a multiple buses configuration. In this case, still referring to three communications, each process has $t_{\text{avail}} = 0$.

Other more sophisticated architectures (e.g., a mesh) can be modeled by setting a proper h_{ave}; in this case, the modeled situation concerns a data flow passing through other components (i.e., bridges or other processors) before the destination is reached; this architectural aspect imposes to add to t_{trans} the delay introduced by each intermediate node.

Finally, it is worth nothing that the last rows of Tables 8.1 and 8.2 refer to communications between different hardware modules (i.e., $hw_i - hw_j$): this characterization is useful in the case of distributed architectures, where the assumption of hw–hw dedicated channels does not apply.

The *Communications Manager*, whose role is to consider the communication issues, exploits the selected parameters during simulation and, with the *scheduler* (i.e., the component implementing the time stretching algorithm), the *Communications Manager* represents the *kernel* of the simulator.

The *Communication Manager*, during the simulation, evaluates t_{avail} and t_{trans} according to both the description and the status of the communication media. Another task of this component is to interact with the

```
If ( n < n_ave )
{
        n ++ ;

        t_trans = s_i * t_bit * f( h_ave, n ) ;

        commList.insert( process i, t_trans ) ;

        ∀ process i in listComm ⇒
                listComm.update(process i, t_transf) ;
}
else
{
        waitQueue.insert( process i ) ;
}
```

Figure 8.5   Starting communications management.

scheduler in order to *wake up* processes that are waiting for communications completion.

The manager adopts a policy described by the algorithms shown in Figures 8.5 and 8.6. The algorithms are based on a list of the communicating processes (*commList*) and a queue of the processes waiting for the communication media (*waitQueue*).

The other parameters are reported below:

- s_i: size, in bit, of the data transmitted by the process i;
- t_{bit}: average time for a bit to cross a single link; and
- $f(h_{ave}, n)$: function of the average number of *hops* and of the number of communications occurred at each step of the simulation. This function provides an expansion coefficient to consider the overhead introduced by routing, control flow algorithms and traffic in the evaluation of t_{trans}.

Anytime a process i initiates a communication, the communication manager performs the actions described by the algorithm reported in Figure 8.5.

If the communication is possible, it starts immediately (the time is managed interacting with the scheduler) and the t_{trans} of the processes in the list are updated to reflect the dependence on the network traffic. If the communication is impossible, the process is inserted in the *waitQueue* until it becomes feasible. The presence within the *waitQueue* implies that the process is in the *idle* state, that is, the process cannot be selected for execution. This waiting implicitly represents t_{avail}. When a communication becomes possible (i.e., one of the current communications terminates), the communication manager follows the algorithm reported in Figure 8.6. If the *waitQueue* is not empty, then the proper

```
If ( waitQueue.isempty() == FALSE )
{
        listComm.insert( waitQueue.extract( process i ), si * tbit * f(have, n) )
}
else
{
        n -- ;
        ∀ process i in listComm
                ⇒ listComm.update(process i, ttrans) ;
}
```

Figure 8.6   Ending communications management.

process is selected, its t_{trans} is evaluated, and it is inserted in the *commList*. Otherwise, the t_{trans} of the communicating processes is updated to reflect the decreasing traffic.

The models and the methodology previously described have been implemented into the *OSTE* co-simulator. The next Paragraph describes in detail the functionalities provided by such a tool.

8.2.2 The Tool

The TOSCA co-simulator (i.e., *OSTE, Occam Simulator for the TOSCA Environment*) has been extended on the basis of the models and the methodology presented above in order to provide heterogeneous multi-processor embedded systems co-simulation capability. Like the other TO(H)SCA tools, it has been developed in C++, and it is based on the *aLICE* library [10].

The TOHSCA co-simulator (always called *OSTE*) supports several steps of the proposed high-level co-design flow (i.e., the *Functional Co-Simulation*, the *Load Estimation* task, and the *Timing Co-Simulation,* Figure 8.1), providing several measures (e.g., profiling, communication costs, etc.) describing the behavior of the whole system.

In the following, the main functionalities provided by this tool are described in detail by focusing on the interaction with the other tools of the flow and on the exchanged data.

Functional co-simulation

The *functional simulation* allows checking the system functionalities to verify their correctness with respect to typical input data sets. This kind of simulation is very fast enabling the designer to easily detect functional errors and anomalous situations like deadlock or the presence of dead code. Moreover, this step provides the following information on the system:

- **Profiling**

For each statement of each procedure composing the system specification, *OSTE* evaluates the average number of executions. This information is very important because it allows to dynamically weighting the different parts of the specification during the following steps.

- **Communication**

For each communication operation (i.e., *procedure calls* and *channel communications*) described in the system specification, *OSTE* evaluates the static cost (i.e., the size of the exchanged data). Such information is used during the *timing co-simulation*, during the *load estimation* task (combined with the profiling), and during the *partitioning* step.

Load estimation

A behavioral system specification, based on an imperative (eventually object-oriented) language, is a powerful entry point for a co-design environment able to support the design of embedded systems. However, when such an approach is used, there is frequently a lack of data able to characterize some important system properties (e.g., procedures execution period, procedure execution times, etc.) and different estimation techniques should be defined. Such properties are instead provided as the starting data set when other forms of specification (i.e., task graph [105]) are adopted. However, this could restrict the co-design environment application field to particular domains (e.g., digital signal applications).

An important property, rarely considered in co-design environments starting from behavioral imperative specifications, is the load that each procedure imposes on a processor-like executor. Since load-balancing issues are very important when dealing with multi-processor systems, an estimation model for the procedures load has been developed and implemented in the co-simulator.

Combining some of the data provided by the previous steps (static timing estimations and profiling) with the designer-imposed timing constraints allows the estimation of the *load* that each OCCAM procedure will impose to a processor (*GPP*) that executes it. The extraction of these data from a behavioral specification is an important task that allows, during the system design exploration step, the evaluation of the number of needed processing elements, the load-balancing task, and the identification of those procedures that will probably need an executor more performing than a GPP.

The intrinsic complexity of such estimation task has lead to the introduction of an assumption that, however, does not invalidate the meaningfulness of the results: for load estimation purpose, each procedure of the specification is assumed periodic.

In this way, the load estimation L_m for $m \in MI$ is provided by the following expression:

$$L_m = \frac{t_m}{T_m} = \frac{t_m \cdot p_m}{T}$$

where t_m is the estimated execution time for m, evaluated with the aim of static timing estimation and profiling considering a mono-processor system, and T_m is the estimated period of m in the hypothesis of periodic execution.

The proposed expression recalls the classical one for the load evaluation, where T_m is obtained by considering the ratio between the designer constraint to the whole execution time and the number of execution of m (the profiling). Therefore, the ratio T/p_m expresses the period that the procedure m would have if it will be periodically executed p_m times in the interval T.

A stronger constraint on T leads to higher L_m and, for each processor in the system, the sum of the loads of the allocated procedures should be less than one, in order to avoid an overloading of the processor that could give rise to the violation of the timing constraint imposed.

Finally, if for a single procedure, L_m is greater than one; this means that the procedure needs an executor more performing than a GPP.

The load estimation task provides to the partitioning tool important information about the behavior of the system, allowing such tool tailoring the system architecture on the basis of the imposed timing constraint considering also load-balancing issues. Such information, together with communication cost, affinity (Chapter 5), and other parameters (Chapter 6), leads to an effective design space exploration.

Timing co-simulation

The *timing co-simulation* considers the heterogeneous multi-processor architecture proposed by the partitioning tool (by means of the *VCG* exchange format, [84]) and a high-level model for the communication media in order to model the system behavior through the behavior of the hardware and software parts. It evaluates the performance of the system verifying its timing correctness.

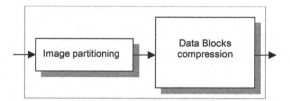

Figure 8.7   The functional composition of the target example.

To this purpose, the simulated system is characterized with the static timing estimations and the affinity (i.e., the VCG graph is annotated), in order to consider in the proper way the procedures allocated on different executors. Moreover, during the timing co-simulation, an analysis is performed on the communication model parameters and the scheduling policy imposed by the designer, in order to detect bottlenecks and suggests possible alternatives. Therefore, this task, validating the solutions proposed by the partitioning and suggesting proper design choices to overcome possible problems, completes the system design exploration step starting from the low-level co-design flow.

8.2.3 Validation

The goal of this Paragraph is to show the effectiveness of the proposed high-level co-simulation methodology when a design space exploration has to be performed. The target application, whose functional decomposition is shown in Figure 8.7, is an abstract representation of a real-time image processing. In this application, the acquired image is preliminary split in blocks of data; such blocks are then processed (e.g., compressed) and forwarded toward other applications. The application *Procedures Call Graph* (Section 3.3.2), shown in Figure 8.8 displays the interactions among the procedures composing the considered application. The *acquire and split* procedure (*Main*) forwards a given data block to two *reservation stations* procedures (P_0, P_1); in turn, each *reservation station* forwards the first data block queued toward one of the four instances of the *data manipulation stations* (C_0, C_1, C_2, C_3). The forwarding operation performed by each *reservation station* has to match with the status of each *data manipulation station*.

Concerning the proposed methodology, a preliminary operation consists of the estimation of the hardware and software execution times for each procedure by using the estimation methodology described in Chapter 6. The estimation methodology allows the extraction of a processor-independent characterization of the software code; such a characterization is linked, in a following

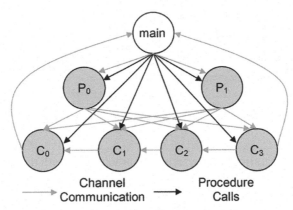

Figure 8.8   The multimedia application: procedure call graph.

phase, to the target processor library. It is worth noting that the proposed methodology allows a fast characterization of the target platform. In the current implementation, the performed analysis is targeted to the *Intel Embedded Ultra Low Power Intel 486 GX* (33 MHz) [76] processor. Concerning the hardware timing characterization, a similar operation is performed; in particular, the timing estimation is technology independent and the actual times are obtained after committing to a technology (in this example a *LSI Logic 10k technology* with a 150 MHz system clock has been used).

The system has been partitioned focusing on the number of processing elements and using metrics concerning both static and dynamic communications. Even if the partitioning step is more articulated, allowing a complex design space analysis (number of processing elements, hardware/software procedure characterization, processor elements type, etc.) for the sake of conciseness the reported analysis has been performed by considering only the number of processing elements. Eight different architectures have been extracted and simulated. The results obtained are shown in Table 8.3.

For each simulated architecture we report, respectively: the number of processing elements, the procedures allocation (the procedures allocated on the same processing element are grouped by brackets), the maximum number of concurrently communications allowed by the interconnection network (in this example h_{ave} is always equal to 0 because the number of processing elements does not require such a feature), the estimated number of clock cycles needed for the application execution by the considered systems, and the simulation time (this time includes the simulator setup time that is always approximately 0.63 s).

By analyzing the results (see Table 8.3), some considerations can be drawn. The less expensive configuration (1 GPP) is also the slowest one. By increasing the number of processing elements, allocation issues must be considered in order to minimize the communication cost and to balance the processors load. It is worth noting that the procedure allocations 2 and 3 in Table 8.3 exhibit different execution times on the same architecture. In the procedure allocation 3, the communication cost is lower than 2 since some of the *reservation station-data manipulation station* data transfers are kept out from the interconnection network and performed locally. Furthermore, in solution 3, the processors load is better balanced since they both execute a single *reservation station task* and two *data manipulation stations* (the load induced by the *Main* procedure is negligible).

Table 8.3 Experimental results

Solution	Architecture	n	$C_{ck}(10^3)$	Simulation Time (s)
1	1 GPP	1	5729	215
2	2 GPP: (M, P_0, P_1) (C_0, C_1, C_2, C_3)	1	4435	255
3	2 GPP: (M, P_0, C_0, C_1) (P_1, C_2, C_3)	1	3289	268
4	3 GPP: (M, P_0, P_1) (C_0, C_1) (C_2, C_3)	1	2877	304
5	5 GPP: (M, P_0, P_1) (C_0) (C_1) (C_2) (C_3)	1	2811	380
6	5 GPP: (M, P_0, P_1) (C_0) (C_1) (C_2) (C_3)	2	2267	350
7	1 GPP: (M, P_0, P_1); 1 HW: (C_0, C_1, C_2, C_3)	1	1446	308
8	1 HW (all-in-hardware)	–	255	103

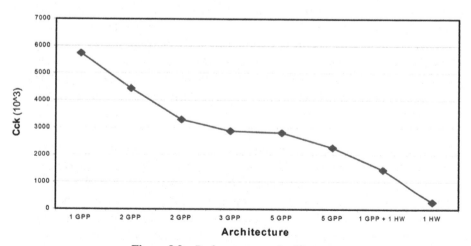

Figure 8.9 Performance vs. Architecture.

By adding other processors, the performance does not increase linearly (Figure 8.9) since the selected interconnection network (i.e., $n = 1$, a single bus) becomes a bottleneck for the communications. In fact, by selecting $n = 2$ (i.e., a crossbar switch or a dual bus configuration), the same 5 GPP architecture gives better results.

Finally, the use of a HW co-processor to execute the *consumers*, or a completely HW implementation, gives a considerably better performance than the other cases (in the *all-in-hardware* configuration, row 8 in Table 8.3, all the communications are performed through dedicated channels).

In general, changing the number and the type of processing elements, the allocation of the procedures, and the features of the interconnection network, different performance estimations (expressed in clock cycles C_{ck}) can be computed. Such a task has been automated and integrated with the partitioning tool, whose goal is also to verify the fulfilling of timing and cost constraints.

8.3 Conclusion

This chapter has introduced a modeling approach and the related simulation strategy, to represent the behavior of multi-processor hardware-software architectures starting from system-level specifications. The simulation kernel has been encapsulated within the TOHSCA co-design toolset and interfaced with the software suite computing the evaluation metrics driving the user during the partitioning task. The proposed approach is particularly valuable since it allows the designer to maintain the analysis at a very abstract level, while gathering significant information on the hypothetical architecture of the final system implementation. The simulator has been implemented in C++ and the experiments have been performed on a SUN *UltraSparc2* running at 256 MHz and equipped with 128 Mbytes *RAM*. The simulator has been able to manage roughly 8000 *events/s* simulating averagely 11000 C_{ck}/s. The number of simulated events per second is particularly relevant whenever the goal of the designer is to verify the system functionality, independently of the target architecture. The second performance figure strongly depends on the characteristic of the target environment and, in this case, it refers to the aforementioned configuration with the microprocessor running at 33 MHz and the hardware components operating at 150 MHz. This level of performance is comparable with other behavioral simulations [56]. The performance of this system-level strategy has been also compared with that of the low-level assembly-like simulation provided by TOSCA (Section 2.1), achieving a speedup of ~ 80, and showing the effectiveness of the system-level approach.

9

Case Studies

The goal of this chapter is to show the applicability and the effectiveness of the proposed flow by describing two case studies. In particular, each step of the flow (shown in Figure 9.1) is considered in detail, describing its role in the environment and the data exchanged with the other tools. Moreover, the case studies focus on the tools used and their interactions in order to emphasize operative issues. Finally, in order to validate the results produced by the proposed methodology and environment, the complete flow has been performed, synthesizing the selected components, and verifying the results by simulation. Section 8.1 applies the complete design methodology to the same example used in Chapter 7 while Section 8.2 presents an example from a real application.

9.1 Case Study 1

The first example considered is the same presented in Section 7.2.3 for the validation of the partitioning tool. It is composed of 52 *OCCAM* procedures and its *Procedure Interaction Graph* (Section 4.3.2) is represented in Figure 9.2. This is not an actual application but it has been obtained by composing some of the procedures belonging to the test suite described in Section 5.2.3. In this way, it has been possible to create an example with a homogeneous distribution of the affinity values. The target architecture is composed of an unconstrained number of *GPP*, *DSP*, and *FPGA*.

In the following, the application of each step of the proposed flow (Figure 9.1) is presented.

9.1.1 Co-specification

The specification of the system is provided by means of the OCCAM specification language. It is a homogeneous specification language: it allows avoiding

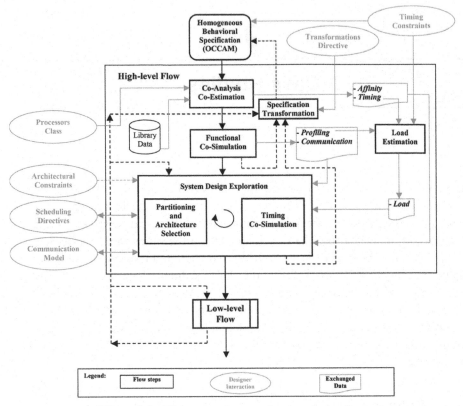

Figure 9.1   The proposed system-level co-design flow.

the detail of possible implementations. In this way, it is possible to perform a complete and unbiased design space exploration.

The *Main* procedure instantiates the main parts of the system and acts as the environment, providing several input data set. Moreover, the specification shows the explicit concurrency in the system represented by the *non-blocking calls* (Section 4.3.2) to *P1*, *P6*, *P9*, *P12*, and *P15*.

9.1.2 Co-analysis

The co-analysis step aims at evaluating the affinity values of each procedure to statically detect the best processing element for the execution (Chapter 5). The *OCCAM analyzer* processes the specification and provides an output file containing a triplet of values for each procedure, i.e., the affinity values

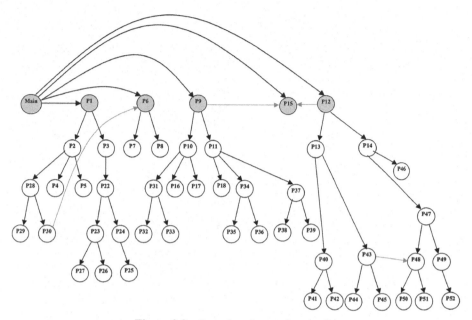

Figure 9.2 Procedure interaction graph.

for *GPP*, *DSP*, and *HW*. Table 9.1 shows the results for the 26 procedures composing the specification. The processing is very fast: the tool, implemented in *C++* and executed on a *Pentium III* running at 700 MHz equipped with 640 Mbytes *RAM*, has processed the OCCAM specification (approximately 1000 LOC) in less than 5 s.

9.1.3 Co-estimation

The next step is very important for *timing co-simulation* purposes: to estimate the performance of the system. The tool *SLET* (Chapter 6) allows the estimation of the clock cycles needed to execute each operation present in the system specification both for software and hardware implementations. The output provided by this tool consists of a file for each procedure. In this example, the performed analysis for the software implementation is targeted to the *Intel 486 GX* [76] processor. The hardware timing characterization has been performed by considering the *LSI Logic 10k technology library* with a 150 MHz system clock. The estimations have been performed on a SUN *UltraSparc2* running at 256 MHz and equipped with 128 Mbytes *RAM*, and the static estimation time has been approximately 2 s.

Table 9.1   Affinity values

Procedure	GPP	DSP	HW	Procedure	GPP	DSP	HW
	Affinity				Affinity		
P1	0.612	0.605	0.397	P27	0.799	0.447	0.492
P2	0.425	0.649	0.388	P28	0.580	0840	0.403
P3	0.535	0.640	0.392	P29	0.460	0.840	0.403
P4	0.259	0.772	0.398	P30	0.559	0.772	0.398
P5	0.297	0.748	0.396	P31	0.868	0.046	0.688
P6	0.543	0.551	0.468	P32	0.597	0.053	0.911
P7	0.658	0.062	0.894	P33	0.997	0.053	0.211
P8	0.997	0.053	0.925	P34	0.643	0.571	0.416
P9	0.586	0.619	0.394	P35	0.580	0.620	0.405
P10	0.548	0.640	0.391	P36	0.544	0.653	0.396
P11	0.351	0.637	0.393	P37	0.608	0.595	0.406
P12	0.623	0.577	0.413	P38	0.520	0.659	0.388
P13	0.553	0.626	0.397	P39	0.553	0.634	0.390
P14	0.424	0.648	0.388	P40	0.604	0.911	0.409
P15	0.524	0.648	0.388	P41	0.471	0.810	0.401
P16	0.997	0.053	0.915	P42	0.352	0.748	0.396
P17	0.612	0.596	0.412	P43	0.459	0.772	0.388
P18	0.587	0.607	0.405	P44	0.259	0.772	0.398
P19	0.616	0.609	0.397	P45	0.574	0.651	0.533
P20	0.725	0.519	0.428	P46	0.384	0.648	0.388
P21	0.565	0.635	0.398	P47	0.524	0.628	0.368
P22	0.997	0.053	0.550	P48	0.234	0.608	0.378
P23	0.658	0.053	0.694	P49	0.987	0.063	0.921
P24	0.799	0.447	0.492	P50	0.992	0.092	0.890
P25	0.494	0.648	0.388	P51	0.968	0.085	0.787
P26	0.384	0.648	0.388	P52	0.989	0.099	0.833

9.1.4 Functional Co-simulation

The functional simulation represents the checkpoint for the functional correctness of the system specification. Moreover, it allows the evaluation of different important measures that characterize the system: profiling, dynamic communication cost, and load estimation. By simulating the system behavior, by means of 10 different typical input data sets (i.e., sequences of approximately 100 values), no dead code and deadlocks have been detected. Moreover, the simulator can provide two output files with the average profiling (i.e., the average number of executions, with respect to the input data sets, of each operation present in the system specification) and the average dynamic communication cost (i.e., profiling multiplied for data size) of each data transfer (i.e., procedure calls and channel communications). The functional simulation has been performed on a SUN *UltraSparc2* running at 256 MHz and equipped with 128 Mbytes *RAM*, and the functional simulation time has been approximately 10 s for each input data set.

9.1.5 Load Estimation

The load estimation task requires the performance estimation data provided by the co-estimation step and a timing simulation. Such a simulation is performed (as described in Chapter 8) by considering a target architecture composed of one general purpose processor that executes all the procedures, thus producing a reference time T_{REF}. Such a task has been performed on a SUN *UltraSparc2* running at 256 MHz and equipped with 128 Mbytes *RAM*, and it has required approximately 10 s for each input data set. The average T_{REF} has been, for the considered architecture, approximately $283 \cdot 10^3$ clock cycles. By imposing as timing constraint to the system an execution time of 40% T_{REF} (i.e., approximately $113 \cdot 10^3$ clock cycles), it has been possible to estimate (as described in Section 8.2.2) the load provided by each procedure to a software executor executing the application with such a constraint. Finally, to represent the information needed for the design space exploration in a compact and general way, a *VCG* [84] file, representing the procedural-level model of the system (Section 4.3.2) annotated with the dynamic communication costs and the estimated load, is provided by the simulator. A small part of such a file is shown in Figure 9.3.

It is worth noting that the following *partitioning and architecture selection* step can be executed based exclusively on such a VCG file and the affinity values, decoupling such a step from the specification language adopted.

9.1.6 System Design Exploration

The system design exploration is divided into two iterative steps: *partitioning and architecture selection* and *timing co-simulation*.

For the partitioning (Chapter 7), based on the considerations reported in Section 7.2.3, the weights of the cost function are set as follows:

$$I_A = I_{Lsw} = I_{Lhw} = 4.0 \quad I_C = I_\$ = 2.0$$

while the ideal load L_{SW} considered in the load index has been set to 70% (a typical value that includes the possible presence of an operating system [141]). The target architecture is composed of an unconstrained number of *GPP*, *DSP*, and *FPGA*.

```
node: { title: "P410" label: "file: occam function: P41 load: 46" }
node: { title: "P420" label: "file: occam function: P42 load: 5" }
node: { title: "P430" label: "file: occam function: P43 load: 80" }
node: { title: "P440" label: "file: occam function: P44 load: 3" }
node: { title: "P450" label: "file: occam function: P45 load: 60" }
node: { title: "P460" label: "file: occam function: P46 load: 5" }
node: { title: "P470" label: "file: occam function: P47 load: 7" }
node: { title: "P480" label: "file: occam function: P48 load: 14" }
node: { title: "P490" label: "file: occam function: P49 load: 2" }
node: { title: "P500" label: "file: occam function: P50 load: 242" }
node: { title: "P510" label: "file: occam function: P51 load: 14" }
node: { title: "P520" label: "file: occam function: P52 load: 2" }
edge: { sourcename: "Main0" targetname: "P10" label: ": communication: 0" }
edge: { sourcename: "Main0" targetname: "P60" label: ": communication: 0" }
edge: { sourcename: "Main0" targetname: "P90" label: ": communication: 0" }
edge: { sourcename: "Main0" targetname: "P120" label: ": communication: 16" }
edge: { sourcename: "Main0" targetname: "P150" label: ": communication: 0" }
edge: { sourcename: "P10" targetname: "P30" label: ": communication: 32" }
edge: { sourcename: "P10" targetname: "P20" label: ": communication: 320" }
edge: { sourcename: "P10" targetname: "P190" label: ": communication: 0" }
edge: { sourcename: "P20" targetname: "P40" label: ": communication: 0" }
edge: { sourcename: "P20" targetname: "P50" label: ": communication: 20" }
edge: { sourcename: "P20" targetname: "P280" label: ": communication: 320" }
edge: { sourcename: "P30" targetname: "P220" label: ": communication: 32" }
```

Figure 9.3 The VCG file.

For the timing co-simulation (Chapter 8), the timing constraint imposed on the execution time of the whole application, as described previously, is 40% T_{REF}. The default policy scheduling is a *round-robin* one and the parameters of the communication model (i.e., number of allowed concurrent communications and number of *hops* [45]) are initially $n_{ave} = 1$ and $h_{ave} = 0$ (i.e., a single bus). The results obtained by applying the methodology described in Section 7.2.1 are shown in Table 9.2.

The iteration 0 considers only one cluster (i.e., only one executor) and the affinity issues suggest the choice of a DSP. However, one executor does not meet the timing constraints. The following iteration 1 found a new minimum for the considered cost function. By simulating such a solution, the simulated time is 51% T_{REF}. The simulator can also provide a measure related to the average number of concurrent communications: in this case, such a number is 1.6 and therefore a new simulation is performed with $n_{ave} = 2$. However,

Table 9.2 Design space exploration

Iteration	I_C	I_{LSW}	I_A	Architecture			Simulated Time	
				GPP	*DSP*	*FPGA*	$n_{ave} = 1$	$n_{ave} = 2$
0	0.000	0.340	0.390	0	1	0	89% T_{REF}	89% T_{REF}
1	0.320	0.040	0.394	3	0	0	51% T_{REF}	45% T_{REF}
3	0.012	0.124	0.386	2	1	0	42% T_{REF}	41% T_{REF}
7	0.006	0.023	0.388	2	1	0	40% T_{REF}	38% T_{REF}

the results obtained are not successful. The same considerations apply in the following of the case study. Another cost function minimum is obtained during iteration 3 (each iteration works with a finer granularity), but also in this case the timing constraint is not satisfied. Finally, iteration 7 reaches the solution: the proposed architecture/allocation (with both the n_{ave} values) satisfies the imposed simulated time.

To perform the design space exploration, the partitioning tool, implemented in $C++$ and executed on a *Pentium III* (700 MHz, 640 Mbytes *RAM*), has processed a population of 700 individuals for 3500 generations in a time less than 20 min (comprehensive of the co-simulation times performed on a *SUN UltraSparc2* running at 256 MHz and equipped with 128 Mbytes *RAM*).

9.1.7 Toward the Low-level Co-design Flow

If the implementation found by the high-level flow is satisfactory, the design process proceeds to the lower levels of the design flow to synthesize the hardware, software, and interface components.

In this example, the first step of the TOSCA low-level flow (Section 3.1) has been applied to show the consistency of the results provided by the high-level one. In particular, the *VIS Compiler* has generated the VIS assembly code considering as target processor the *Intel 486 GX* [76] for the procedures allocated on the *GPP*, and the *Analog Devices SHARC DSP* [91] for the procedures allocated on DSP. The evaluation of the clock cycles needed to execute each assembly instruction has been performed based on the technology libraries provided by the processors vendors and the results have been backannotated to the related OCCAM source code. Then, a new system-level timing co-simulation has been performed to check the consistency of the data provided in the last step of the high-level flow, by adopting the back-annotated data instead of the system-level estimation ones (Chapter 4). The simulations, based on the same input data set, have provided simulated times belonging to the range 37–48% T_{REF}, thus substantially similar to the previous ones.

9.2 Case Study 2

The second case study is related to an application that manages digital signatures in MPEG encoded images. The application is described by means of 21 *OCCAM* procedures and its *Procedure Interaction Graph* (Section 4.3.2) is shown in Figure 9.4. The target architecture is composed of an unconstrained

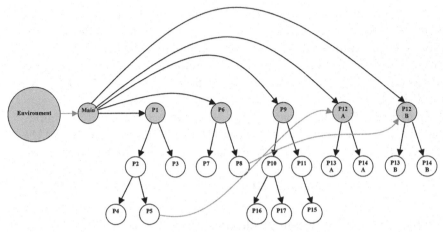

Figure 9.4   Procedure interaction graph.

number of *GPP*, *DSP*, and *FPGA*. In the following, each step of the application of the proposed flow (Figure 9.1) is presented.

9.2.1 Co-specification

The specification of the system is provided by means of the OCCAM specification language. The *Main* procedure instantiates the main parts of the system and interacts with the procedures that represent the environment providing several input data sets (i.e., MPEG files). Moreover, the specification shows the explicit concurrency in the system represented by the *non-blocking calls* (Section 4.3.2) to *P1*, *P6*, *P9*, and the two instances of *P12*.

9.2.2 Co-analysis

The first step aims at evaluating the affinity values of each procedure to statically detect the best processing element for their execution (Chapter 6). The *OCCAM analyzer* processes the specification and provides an output file containing a triplet of values for each procedure, as shown in Table 9.3 where, for each procedure, the *GPP*, *DSP*, and *HW* affinity values are reported. The process is very fast: the tool, implemented in *C++* and executed on a *Pentium III* running at 700 MHz equipped with 640 Mbytes *RAM*, has processed the OCCAM specification (approximately 1600 LOC) in less than 10 seconds.

Table 9.3   Affinity values

Procedure	Affinity		
	GPP	DSP	HW
P1: fdct	0.833	0.505	0.297
P2: get_HDR	0.417	0.429	0.188
P3: init_fdct	0.235	0.740	0.352
P4: init_buffer	0.259	0.772	0.398
P5: fill_buffer	0.879	0.248	0.796
P6: decode	0.542	0.751	0.448
P7: next_start_code	0.785	0.126	0.294
P8: detect_sequence_header	0.297	0.853	0.925
P9: copyright_extension	0.856	0.691	0.294
P10: macroblock_modes	0.848	0.240	0.193
P11: motion_compensation	0.311	0.667	0.394
P12: check_frame	0.223	0.577	0.314
P13: subsample_vertical	0.153	0.624	0.290
P14: subsample_horizontal	0.124	0.647	0.382
P15: motion_vectors	0.524	0.848	0.138
P16: add_block	0.977	0.253	0.351
P17: decode_SNR_block	0.312	0.696	0.212

9.2.3 Co-estimation

The next step estimates the performance of the system. The tool *SLET* (Chapter 6) allows the estimation of the clock cycles needed to execute each operation present in the system specification both for software and hardware implementations. The output provided by this tool consists of a file for each procedure. In this example, the performed analysis is targeted to the *Intel 486 GX* [76] processor. The hardware timing characterization has been obtained considering the *LSI Logic 10k technology library* with a 150 MHz system clock. The estimations have been performed on a SUN *UltraSparc2* running at 256 MHz and equipped with 128 Mbytes *RAM*, and the static estimation time has been approximately 3 s.

9.2.4 Functional Co-simulation

The functional simulation determines the functional correctness of the system specification. Moreover, it allows the evaluations of important parameters characterizing the system: profiling, dynamic communication cost, and load estimation. After simulating the system behavior by means of four different typical input data sets (i.e., MPEG coded files), no dead code and deadlocks have been identified. Moreover, the simulator provides two output files with the average profiling (i.e., the average number of executions, with respect to the input data sets, of each operation present in the system specification) and the average dynamic communication cost (i.e., profiling multiplied by data size) of each data transfer (i.e., procedure calls and channel communications). The functional simulation has been performed on a SUN *UltraSparc2* running at 256 MHz and equipped with 128 Mbytes *RAM*, and the functional simulation time has been approximately 30 s for each input data set.

9.2.5 Load Estimation

The load estimation task requires the performance estimation data provided by the co-estimation step and a timing simulation. Such simulation is performed (as described in Chapter 8) considering a target architecture composed of one general purpose processor that executes all the procedures, thus producing a reference time T_{REF}. Such a task has been performed on a SUN *UltraSparc2* running at 256 MHz and equipped with 128 Mbytes *RAM*, and it has required approximately 1 min for each input data set. The average T_{REF} has been, for the considered architecture, approximately $283 \cdot 10^6$ clock cycles. By imposing as timing constraint to the system an execution time of 50% T_{REF}, it has been possible to estimate (as described in Section 8.2.2) the load provided by each procedure on a software executor executing the application with such a constraint. Finally, to represent the information needed for the design space exploration in a compact way, a *VCG* file [84], representing the procedural-level model of the system (Section 4.3.2) annotated with the dynamic communication costs and the estimated load, is provided by the simulator. It is worth noting that the following *partitioning and architecture selection* step can be executed based exclusively on such a VCG file and the affinity values, decoupling such a step from the specification language adopted.

9.2.6 System Design Exploration

The system design exploration is divided into two iterative steps: *partitioning and architecture selection* and *timing co-simulation*.

The weights associated with each parameter constituting the cost function used by the partitioning tool to identify the best solution have been set to:

$$I_{\text{A}} = I_{\text{Lsw}} = I_{\text{Lhw}} = 4.0 \qquad I_{\text{C}} = I_{\$} = 2.0$$

while the ideal load L_{SW} considered in the load index has been set to 70% (a typical value that includes the possible presence of an operating system [141]). The target architecture is composed of an unconstrained number of *GPP*, *DSP*, and *FPGA*.

About timing co-simulation (Chapter 8), the timing constraint imposed on the execution time of the whole application, as described previously, is 50% T_{REF}; the default policy scheduling is a *round-robin* one and the parameters of the communication model (i.e., the number of allowed concurrent communications and the number of *hops* [45]) are initially $n_{ave} = 1$ and $h_{ave} = 0$ (i.e., a single bus). The results obtained by applying the methodology described in 7.2.1 are shown in Table 9.4 and described in the following.

The iteration 0 considers only one cluster (i.e., only one executor) and the affinity issues suggest the choice of a DSP. However, one executor does not meet the timing constraints. The following iteration 1 found a new minimum for the considered cost function. Simulating such a solution, the simulated time is 71% T_{REF}. The simulator can also provide a measure related to the average number of concurrent communications: in this case, such a number is 1.85 and so a new simulation is performed with $n_{ave} = 2$. However, the results obtained are not successful. The same considerations apply in the following of the case study. Another cost function minimum is obtained during iteration 6 (each iteration works with a finer granularity) but also in this case the timing constraint is not satisfied. Iteration 8 identifies a solution: the proposed architecture/allocation with $n_{ave} = 2$ (e.g., a crossbar switch) reaches the imposed simulated time; however, the margin is considered too low and so another iteration is performed. The last iteration, changing only the allocation of the procedures on the same architecture, finds a better solution that is considered acceptable. To perform such a design space exploration, the partitioning tool, implemented in *C++* and executed on a *Pentium III* (700 MHz, 640 Mbytes *RAM*), has processed a population of 500 individuals

Table 9.4   Design space exploration

Iteration	I_C	I_{LSW}	I_A	Architecture GPP	DSP	FPGA	Simulated Time $n_{ave} = 1$	$n_{ave} = 2$
0	0.000	0.241	0.292	0	1	0	92% T_{REF}	92% T_{REF}
1	0.120	0.140	0.344	0	2	0	71% T_{REF}	65% T_{REF}
6	0.201	0.64	0.376	1	1	1	58% T_{REF}	52% T_{REF}
8	0.146	0.73	0.398	1	2	0	54% T_{REF}	49% T_{REF}
9	0.138	0.72	0.401	1	2	0	50% T_{REF}	47% T_{REF}

for 3500 generations in a time less than 35 min (comprehensive of the co-simulation times performed on a *SUN UltraSparc2* running at 256 MHz and equipped with 128 Mbytes *RAM*).

9.2.7 Toward a Low-level Co-design Flow

If the implementation found by the high-level flow is satisfactory, the design process proceeds to the lower levels of the design flow. In this example, the first step of the TOSCA low-level flow (Section 3.1) has been applied to check the consistency of the results provided by the high-level exploration. In particular, the *VIS Compiler* has generated the placeVIS assembly code considering as target processor the *Intel 486 GX* [76] for the procedures allocated on the *GPP*, and the *Analog Devices SHARC DSP* [91] for the procedures allocated on the DSP. The evaluation of the clock cycles needed to execute each assembly instruction has been obtained based on the technology libraries provided by the processors vendors and the results have been backannotated to the related OCCAM source code. Then, a new system-level timing co-simulation has been performed to check the consistency of the data provided in the last step of the high-level flow, adopting the back-annotated data instead of the system-level estimation ones (Chapter 4). The simulations, based on the same input data set, have provided simulated times belonging to the interval 42–55% T_{REF}, substantially confirming the previous ones.

9.3 Conclusion

This Chapter has shown the applicability and the effectiveness of the proposed flow by describing two case studies. In particular, each step of the flow has been considered in detail, describing its role in the environment and the data exchanged with the other tools. Moreover, the case studies have focused on the tools used and their interaction in order to emphasize the operative issues. Finally, a first step toward the low-level co-design flow is considered showing the consistency of the provided results.

Conclusions (Part 1)

The main contribution of the first part of the book (i.e., Part 1) is the definition of an innovative approach to the problem of hardware/software co-design of heterogeneous multi-processor embedded systems, considering in particular the design space exploration phase in terms of target architecture and hw/sw partitioning.

Multi-processor embedded systems are a promising solution for a broad range of modern and complex applications. However, their design complexity is relevant, and no assessed design methodology is available today. Hence, a systematic approach to the problem has been developed considering both theoretical aspects and pragmatic issues, resulting in a consistent set of models, a complete methodology, and a toolset able to support the system-level concurrent design of embedded systems, possibly under real-time constraints. A co-design environment has been developed to support the designer from the specification phase at a high level of abstraction, to the definition of the target architecture and the identification of the best hw/sw partitioning. Furthermore, the tools developed provide directives to drive the actual synthesis and co-simulation phases, that can be performed with existing commercial or prototype tools (tools for high-level synthesis, compilers, low-level co-simulators, etc.).

The input specification describes the overall functionality of the system and includes the desired constraints. In our case, the entire design flow is based on the *OCCAM2* language for which almost no tools are publicly available. For this reason, the tools developed for the existing *TOSCA* project (parser, compiler, linker, several libraries, etc.) have been integrated in the new environment.

Starting from such a specification, the developed environment supports automatic and/or interactive exploration of the design space, which leads to the identification of high performance and cost-effective solutions.

The final goal has been to contribute to the extension of an existing single processor co-design environment (i.e., *TOSCA*) enabling it to support co-specification, co-analysis, co-estimation, co-verification, and system design exploration of heterogeneous multi-processor architectures at system-level. In particular, the research has produced the following main contributions, covering all the main steps of the high-level co-design flow, with respect to the state of the art. An internal model has been developed to represent the OCCAM2 specification language. Based on such a model, a set of innovative metrics has been defined. Such metrics, allow the co-analysis step to identify and evaluate the functional and structural properties of the specification, which could affect design performance on different architectural platforms. Moreover, a model and a methodology to estimate software and hardware performance at a high-level have been proposed. The information provided by the co-analysis and co-estimation step allows an effective design space exploration, based on two main tasks: *partitioning and architecture selection* and *timing co-simulation.*

The partitioning methodology is based on an initial *clustering* and on a heuristic optimization algorithm. Such an algorithm provides solutions that consist of an architecture and the binding between parts of the system behavior and the selected components. Such solutions are then validated by means of a system-level co-simulation strategy that considers the presence of multiple executors and a high-level model for the communications.

The experimental results obtained are encouraging and justify the continuation of the research in this direction. The current efforts are twofold: to apply the presented approach to *SystemC*-based environments, extending the existing methodologies and tools and integrating such tools with those presented in this book; to validate the whole flow by means of measures performed on actual prototypes.

In particular, the migration to the SystemC is based primarily on the generality of the procedural-level internal model developed in this work, while for the co-analysis and co-estimation methodologies work is in progress to apply the developed models to such a language. In this way, it will be possible to build an annotated *VCG graph* representing the main features of the system enabling the use of the presented partitioning methodology and thus, by means of the existing SystemC co-simulator, properly integrated with a high-level model for the communications, it will be possible to validate the proposed solutions.

Finally, in order to perform a complete validation, the proposed system-level flow will be integrated with existing synthesis tools in order to derive prototypes based on a innovative target architecture composed of several processors and an FPGA integrated on a single chip.

Part 2

January 2002–August 2014

10

System-Level Design Space Exploration for Dedicated Heterogeneous Multi-Processor Systems

10.1 Introduction

Systems based on heterogeneous multi-processor architectures (i.e., HMPS, *Heterogeneous Multi-Processor Systems*) have been recently exploited for a wide range of application domains, especially in the *System-on-Chip* (SoC) form factor (e.g., [143–146]). Such systems include several processors, memories, and a set of interconnections between them. By definition, the set of processors in the same is heterogeneous. This implies that it is possible to have (following a slight variation of the HW/SW unifying classification proposed in [158]), at the same time, one or more processors belonging to different classes:

- COTS (i.e., *Common-Off-The-Shelf*) *general-purpose* processors (GPP, e.g., *ARM*, *MIPS*, *MicroBlaze*, Nios II, etc.)
  - Their main feature is to be able to execute a fixed and standard *Instruction Set Architecture* (ISA)
- COTS *domain-oriented* processors (e.g., DSP, *Digital Signal Processor*; GPU, *Graphical Processing Unit*; *etc.*)
  - Their main feature is to be able to execute a fixed and domain-specific ISA
- Custom *domain-oriented* processors (the so-called ASIP, *Application Specific Instruction Processor*)
  - Their main feature is to be able to execute a customizable domain-specific ISA
- COTS *single-purpose* (or *specific-purpose*) processors (SPP, e.g., *AES coder*, *JPEG coder*, *UART/SPI/I^{2}C Controller*, etc.)

○ Their main feature is to be able to execute a standard specific function (i.e., no ISA involved)

• Custom *single-purpose* (or *specific-purpose*) processor (SPP, i.e., the actual ad-hoc developed digital HW component)

   ○ Their main feature is to be able to execute a custom-specific function (i.e., no ISA involved)

Such processors can be adopted in the form of (*soft*, *hard* or *fuse*) *IP cores* or as *discrete ICs* mainly depending on the final system form factor (i.e., *on-chip*, *on-FPGA*, *on-board*) and scope (final product or platform).

As stated in the title, this Chapter focuses on *dedicated systems*, i.e., a digital electronic system with an HW/SW custom architecture that is specifically designed in order to satisfy *a priori known* application requirements (both functional and not functional). A dedicated system could be *embedded* in a more complex system or it could be subjected to *real-time* constraints. Often, both the previous situations can apply.

When dedicated systems are also HMPS (i.e., D-HMPS) implemented by means of a meaningful (i.e., hundreds) number of *processors* in a single ICs, they are also called *Dedicated Heterogeneous Multi-Processor Systems*. Apart from possible differences on terminology and composition, for this kind of architectures one consideration is always true: they are so complex that the *co-design methodology*, i.e., the set of adopted models, metrics, and tools and the whole organization of the design activities (a.k.a. the *co-design flow*), plays a major role in determining the success of a product. In fact, in the past years, a remarkable number of research works have focused on system-level co-design of HMPS (e.g., [96, 121, 136, 147–156], and [157] for a meaningful analysis of the issue). Each of them has proposed a different approach to the design space exploration approach but all of them always rely on a fixed target HW architecture or heavily rely on the designer's experience to define some of the target HW architecture features. In particular, the definition of the communication architecture (i.e., the details of the interconnection links and the topology) is always only an input (typically imposed by a platform-based approach) to the co-design flow. So, up to now, there does not exist a system-level co-design flow that fully addresses the problem of automatically suggesting an HW/SW partitioning of the system functionalities specification while also mapping the partitioned entities onto an automatically selected heterogeneous multi-processor architecture considering both computation and communication issues. In the historic literature, a special mention is due to the inspiring work presented so far in [160], where the main difference with the presented one is related to the need of a pre-defined architecture and some binding directives that should be provided by the designer. However, such a work takes into consideration also the scheduling issues while the presented one considers them only in a partial way. For this, the scope of the present Chapter is limited to *system-level design space exploration* instead of *system-level synthesis* as in [160].

According to this scenario, this Chapters proposes the arrival point of the research path that has started from the work described in Part 1 has reached a meaningful milestone in [150] and has been partially extended in [166]. The final goal is the definition of a design space exploration approach that, starting from the system functionalities specification and related requirements, would be able to suggest to the designer:

- an HW/SW partitioning of the given system functionalities specification;
- a D-HMPS architecture (considering both computation and communication elements); and
- a mapping of the partitioned entities onto the proposed architecture able to satisfy imposed requirements.

The approach described in the following paragraphs proposes a whole re-organization of the main concepts aiming to overcome previous limitations and to introduce some innovations by means of a refined abstract modeling strategy able to preserve the general applicability of the methodology and the feasibility of the proposed solutions.

10.2 Reference Co-Design Flow

The reference system-level co-design flow (a slight extension of the one described along Part 1) is shown in Figure 10.1: it reports the organization of the main design activities that are briefly described below.

Specification

The entry point of the proposed co-design flow is the specification of the desired system functionalities with related timing constraints. A key point is the adoption of a homogeneous specification language to avoid polarizing the design toward hardware or software at this early stage of the flow. More details about the adopted model of computation and specification languages are provided in Paragraph 10.3.

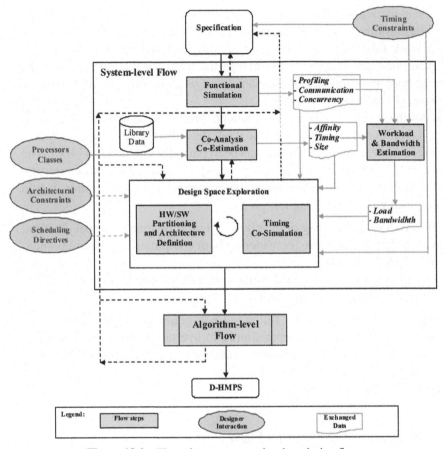

Figure 10.1   The reference system-level co-design flow.

Functional Simulation

The first step is the *Functional Simulation* (Chapter 8, [161]), where the system functionalities are simulated to check their correctness with respect to typical input data sets and related outputs. This kind of simulation is very fast and allows the designer to detect functional errors but it is strictly dependent on the meaningfulness of selected input data set. The early detection of anomalous behaviors allows the designer to correct the specification avoiding a late discovery of problems that could lead to time-consuming design loops. During this step, important data characterizing the dynamic behavior of the system are also collected: *Profiling, Communication,* and *Concurrency.* This means, respectively, that it is possible to evaluate min/max number of executions of each statement composing the specification, the min/max amount of data exchanged between different functionalities of the specification, and to identify the functionalities that are executed concurrently.

Co-Analysis and Co-Estimation

This step of the flow aims at extracting as much information as possible about the system by statically analyzing the specification. This step is composed of the *Co-Analysis* (Chapter 5, [150, 164]) and *Co-Estimation* (Chapter 6, [167–169]) activities. The former one provides a set of data expressing the *Affinity* of the system functionalities toward a set of processor classes (actually this work considers only COTS GPP, COTS DSP, and Custom SPP), while the latter provides a set of estimations of the *Timing* required by each processor class in the set for the execution of each single statement that composes the specification. The timing estimation tool requires more details about the processors class that will execute the code (i.e., x86, ARM, etc.), and a data library associated with the selected class (obtainable from processor documentation) and to the selected hardware technologies (i.e., FPGA family). It is worth noting that the estimation tool is dependent on the processors class only through such a library and so it can be used potentially for every kind of processor. Finally, another important estimation is related to the *Size*: min/max ROM and RAM bytes needed for SW implementations and min/max *equivalent gate* (or equivalent metrics such as the number of *cells* or *LUTs*) for HW ones.

Load and Bandwidth Estimation

Combining the data provided by the previous steps (timing and profiling data) under a *time-to-completion* constraint allows the estimation of the min/average/max *Load* (Chapter 8) associated with the execution of each system functionality on a single COTS GPP system (i.e., typically the worst case). The extraction of these data from the system specification is an important task that allows, during the system design exploration step, the evaluation of the number of needed processors and the identification of those functionalities that probably need a processor more performing than a GPP to satisfy a time-to-completion timing constraint. Finally, combining communication data with timing data, it is possible to estimate the min/max *Bandwidth* needed to the different functionalities to exchange data while fulfilling a time-to-completion constraint.

Design Space Exploration

Finally, the flow reaches the *Design Space Exploration* step that is constituted by two iterative tasks: *HW/SW Partitioning and Architecture Definition* (Chapter 7, [150]), and *Timing Co-Simulation* (Chapter 8, [161, 162]). All the data produced in the previous steps are used to guide the process, together with additional information provided by the designer. Such information expresses the *Architectural Constraints* and the *Scheduling Directives* (i.e., available scheduling policies and possible priorities among system functionalities). The partitioning and architecture definition task explores the design space looking for feasible mapping/architecture items suitable to satisfy imposed constraints. Then, the timing co-simulation task will consider the suggested mapping/architecture items to actually check for timing constraints satisfaction. If the proposed mapping/architecture item does not meet some timing constraints, the designer could perform again the design space exploration by changing some exploration parameters or by modifying the system functionalities specification.

Specification Transformation

This implicit step (i.e., no related box in Figure 10.1) involves the specification transformation that can be performed in order to help satisfying the constraints. In detail, it is possible to go back in the flow from several points (dotted lines in Figure 10.1), each of them more costly than the previous one, that is: after the functional co-simulation; after the design space exploration; and after the algorithm-level flow (described in the following). The first situation arises when the functional simulation helps to detect some functional errors in the system behavior, that is, there are some functional errors in the system functionalities specification itself. In this case, the designer is called to correctly express the desired system behavior. In the second case, the design space exploration step is not able to provide a mapping/architecture item able to satisfy the constraints. To solve this problem, other than trying to change design space exploration parameters or architectural constraints and scheduling directives, it is possible to modify the system functionalities specification by applying some transformations suitable to explicitly show some system features (e.g., concurrency) that the tools could take into account and better exploit in the following steps. An historic meaningful example of such an approach could be found in [170], where the exploitation of the formal nature of the OCCAM language (i.e., based on *process algebra*) allowed for automatic transformations providing several semantically equivalent OCCAM specifications to be used as entry points for an HW/SW partitioning tool. If such modifications do not provide any successful response, one or more constraints should be eventually relaxed or the system is not feasible with the selected HW/SW technologies. Finally, there is the case in which after the algorithm-level flow the obtained solution does not satisfy some of the constraints. This is the worst situation and the goal of a co-design methodology is to avoid it. However, this possibility should be taken into account and so there are different possible actions: to re-execute the algorithm-level flow trying to change its parameters, to re-execute the design space exploration trying to change its parameters, to modify the specification, to relax some constraints, or to change selected HW/SW technologies.

Algorithm-Level Flow

When the mapping/architecture item proposed by the design space exploration is satisfactory, there is the need to implement the system by means of suggested processors, memories, and interconnections. For this, the system functionalities allocated on general-purpose or domain-oriented processors will be typically transformed in C code with the support of a possible embedded and/or real-time OS, while the system functionalities allocated on single-purpose processors will be typically transformed in synthesizable HDL code or implemented by means of existing COTS component. It is worth noting that the cited transformations will be done automatically or manually depending on language and style adopted to write specification. Finally, depending on the final system form factor (i.e., *on-chip*, *on-FPGA*, *on-board*), a different approach will be needed to physically obtain processors, memories, interconnections, and HW/SW interfaces. Such approaches are all related to existing commercial algorithm-level methodologies and tools (Figure 10.2 show a typical algorithm-level design flow for *on-chip* and *on-FPGA* approaches) that are out of the scope of this Chapter.

10.3 Specification

The entry point of the proposed co-design flow is the specification of the desired system functionalities and the first step of the flow aims at extracting as much information as possible about the system by analyzing such a specification. These aspects introduce some *modeling issues* that imply some choices about the adopted model of computation (*MoC*) and related specification languages to be used to describe the system functionalities, and about the

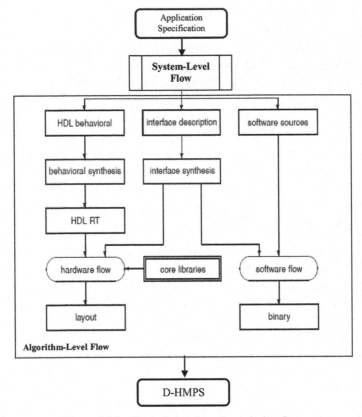

Figure 10.2   Typical algorithm-level design flow.

"internal models" of representation to be used in order to allow a proper tool-chain to make automatic analysis and transformations.

In this work, the system functionalities specification is assumed to be based on the CSP model of computation (*Communicating Sequential Processes* [37, 159]) and described by means of an homogeneous (imperative or OO) language suitable to support such an MoC (e.g., *OCCAM* [12], *HandelC* [184], *ImpulseC* [172], *SystemC* [14], etc.). This choice is based on the common belief (e.g., [157, 158]) that such an MoC is very suitable for the system level and it is also very flexible. In fact, the communication scheme used by the CSP (i.e., unidirectional blocking channels) could be easily adapted to different needs by introducing additional processes: e.g., if two processes need an asynchronous communication scheme, it would be sufficient to introduce a *buffer process* between them [37, 159].

Once selected, an MoC, the choice of a specification language is mainly related to convenience: it should be easy to describe the system functionalities, and there should exist a proper support of documentation and tools (i.e., parsers, simulators, etc.). For example, OCCAM is based on CSP and it is theoretically very suitable but, apart from some academic use (e.g., the flow described in Part 1 of this book), it is difficult to find adoptions in real-world co-design flows. HandelC, strictly derived for OCCAM, is another suitable language that has been already proposed in commercial methodology and tool (e.g., [171]). However, it is more oriented to HW than to HW/SW system. Another approach is offered by ImpulseC, but it is again more oriented to HW and its use in a system-level context should be still investigated. Finally, one of the most promising languages is SystemC because it deals with real system-level concepts (i.e., *Transaction-Level Modeling* [189]) and it is very well supported. Moreover, it is also straightforward to represent CSP by means of it (e.g., [173–175]).

Once selected, an MoC and a language, in order to automatically analyze the specification, two *internal models* at different levels of abstraction are needed:

- a *statement-level* internal model, that strictly depends on the chosen specification language and on existing parsers, used to analyze the specification at an abstraction level detailed enough to make reliable estimations and to compute metrics; and
- a *procedure-level* internal model, possibly independent of specification languages, used to explore the design space.

Since the main internal model for the design space exploration is the procedure-level one, the approach presented in this Chapter is based on an internal model called *Procedural Interaction Graph* (*PING*) (Chapter 4, [150]). The *Procedural Interaction Graph* is a formalism that provides information about the relationships among procedures and the exchanged data. Such a graph is suitable for representing a coarse grain view of system functionalities that takes into account communications, synchronizations, and concurrency issues. An intuitive graphical representation of such a graph (Figure 10.3) is obtained by associating a node with each instance of a function/method (the node is gray filled if the function/method is non-blocking) and an arrow for each data transfer (the arrow is black for procedure/method calls, while it is gray for communication/synchronization operations). By means of a PING, it is possible to represent a CSP while taking into account also possible decomposition of each CSP process in classical functions/methods.

Figure 10.3 graphically shows a CSP (with 6 processes and 6 channels) and a possible related *PING* where the processes have been decomposed in simpler elements (i.e., procedures). For example, process P1 in CSP is decomposed in procedures P1, P1.1, and P1.2, process P2 is stand-alone, and processes P3 and P4 are decomposed by adding one procedure for each one (i.e., P3.1 and P4.1). The dotted *Main* procedure is fictitious and it is needed only to consider possible instantiation issues. It is worth noting that non-blocking calls have been used to reflect the potential concurrent nature of the sets {P1, P1.1, P1.2}, {P2}, {P3, P3.1} and {P4, P4.1} as already expressed in CSP. Moreover, the blocking calls explicitly reflect that inside each set, no concurrency is possible. Finally, the decomposition allows also to specify which parts of processes (i.e., procedures) are really involved in data exchange. For example, it is possible to specify procedure calls with possible parameters exchange and return value (e.g., P1 and P1.1) and the procedures related to the *channels* expressed in the CSP (e.g., channel *ch2* is related to P1.1 an P2). In such a sense, a PING is able to explicitly show, with a procedure-level coarse grain, concurrency and communication/synchronization issues.

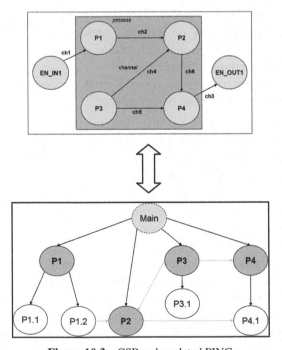

Figure 10.3   CSP and a related PING.

It is worth noting that the designer can proceed in the specification by means of the following approach. First, the system functionalities are decomposed in processes in order to explicitly identify concurrency and communication. Then, each process is decomposed in procedures to explicitly identify related block of statements that could be grouped together and to refine communication needs. Finally, by means of the selected specification language, each procedure in the PING is fully described. So, from now on, in order to clarify the terminology used in previous sections, a single procedure in the PING is supposed to be a desired system functionality. To end the specification step, there could be the need to add timing constraints to the specification. This can be done by exploiting the features of the adopted specification language or by extending it as done, for example, with the OCCAM language in Chapter 4. Basically, there could be the need to provide, at least, some mechanisms to specify min/max delay or execution rates (as described so far in [176, 177]) for the execution of a group of statements, a PING procedure, or a group of PING procedure.

10.4 Target HW Architecture

The target HW architecture selected for the proposed methodology is a heterogeneous multiprocessor one with distributed local memory. So, it is composed of proper interconnections of some instances of different basic elements. The basic elements, called *Basic Block* (BB, Figure 10.4), represent the minimal computation, storage, and communication units in the system. They can be different in their internal components giving so rise to possible heterogeneous multiprocessor systems. In particular, each BB is composed of three main elements: a *Processing Unit* (PU), a *Local Memory* (LM), and an *External Communication Unit* (ECU). These elements are interconnected by an *Internal Interconnections Link* (*IIL*, typically a shared bus where PU is the master and ECU is a memory-mapped SPP slave).

In the actual version of the methodology, PU could belong only to one of three different processor classes (i.e., COTS GPP, COTS DSP and Custom SPP), where GPP and DSP are characterized by the cost (€) and the maximum load (L_{MAX}, that could be kept less than 100% to take into account a possible OS or to satisfy possible WCET constraints), while SPP is characterized by the cost (€) and the maximum number of available equivalent gates $G_{\mathrm{eq\_MAX}}$ (in the case of a fixed family of reconfigurable logic, the last metric has to be changed with the max number of available *cells* or *LUT*). LM is the local memory directly addressable by the PU (i.e., no CSP channels are involved

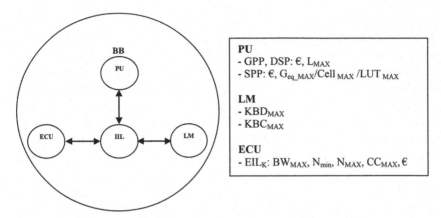

Figure 10.4   The Basic Block and its characterization parameters.

for this kind of communications) and it is characterized by a cost (€) and max size for data (KBD_{max}) and max size for code and parameters (KBC_{max}). Finally, ECU is characterized by the set of *External Interconnection Links* (*EIL*) that it is able to manage. Each EIL is further characterized by the following parameters:

- the max available bandwidth (BW_{max});
- the min/max number of BBs that shall/can use a single EIL instance (N_{min} and N_{max});
- the max number of allowed concurrent communications (CC_{max}); and
- the cost (€).

So, given some instances of BBs and interconnecting them by means of some instances of EILs (it is worth noting that a pair of CUs should be able to manage at least a common EIL in order to allow the related BBs to communicate), it is possible to define a feasible dedicated heterogeneous multiprocessor architecture on which the system functionalities can be mapped. Such an architecture can be then represented by means of an architecture graph [160] (the same notation has been used for the BB in Figure 10.4).

For example, Figure 10.5 shows an architecture graph composed of 6 instances of BBs where the PUs are DSP (named *DSP1*, i.e., the instance number 1 of the DSP processor class, and *DSP2*) for *BB1* and *BB6*, GPP (named *GPP1, GPP2,* and *GPP3*) for *BB3, BB4,* and *BB5*, and SPP (named *SPP1*) for *BB2*. Such BBs are connected by means of three types of

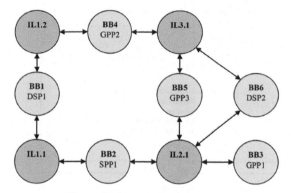

Figure 10.5 Architecture graph.

interconnections links (the *external* prefix will be avoided until the end of the Chapter) *IL1*, *IL2*, and *IL3*. The instance number 1 of IL1 (i.e., *IL1.1*) connects BB1 and BB2, *IL1.2* connects BB1 and BB4, *IL2.1* connects BB2, BB3, BB5, and BB6, and, finally *IL3.1* connects BB4, BB5, and BB6.

10.5 Design Space Exploration

This section explains in detail the design space exploration approach proposed for the "HW/SW Partitioning and Architecture Definition" task of Figure 10.1, with the adopted heuristics, metrics, and cost functions. The final goal is the automatic identification of an HW/SW partitioning of the system functionalities (i.e., the PING procedures), a D-HMPS architecture, and a mapping of the partitioned entities onto the architecture able to optimize the adopted cost functions. The proposed approach is decomposed into two sequential phases, as shown in Figure 10.6.

10.5.1 First Phase

The first phase is related to the mapping of PING procedures onto a dedicated architecture while considering only computation issues (based on the methodology and an enhancement of the tools described in Part 1 and [150]). The internal-model representing the specification, annotated by means of the *Co-Analisys & Co-Estimation* step (Figure 10.1), is provided as input to the *PAM1* (i.e., *Partitioning, Architecture Definition and Mapping Phase 1*) tool. In particular, each procedure in the PING is annotated with several metrics:

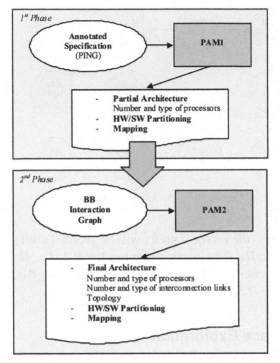

Figure 10.6   The two-phase DSE approach.

- min/max load imposed by each procedure to a single COTS GPP under a time-to-completion constraint: l;
- min/max bandwidth needed to communicate with other procedures while fulfilling a time-to-completion constraint: b;
- min/max size for HW and SW implementations: s (*KBD* and *KBC* bytes, and G_{eq}/*cells*/*LUT*);
- affinity of each procedure toward a set of processor classes (actually only COTS GPP, COTS DSP, and Custom SPP): a (it is worth noting that the affinity values could be also provided manually by the designer in the case he would like to exploit his experience).

So, starting from an annotated PING (Figure 10.7 shows an annotated version of the one in Figure 10.3), the first phase goal is to determine number and type of BB/PUs and a mapping of PING procedures onto them trying to:

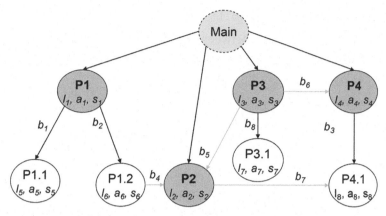

Figure 10.7 Procedural interaction graph.

- minimize the cost of the set of BBs
  - The max number of instances for each kind of PUs could be provided as an architectural constraint by the designer (to limit cost, to ensure feasibility, or to model an existing platform)
- keep the load of each PU near but under its L_{MAX}
- minimize communications between different BBs
- exploit to the best the affinity between PUs and the mapped procedures
- keep the used size near but under $\mathrm{KBD_{MAX}}$, $\mathrm{KBC_{MAX}}$ (for GPP/DSP), or $G_{\mathrm{eq\_MAX}}/\mathrm{Cell_{MAX}}/\mathrm{LUT_{MAX}}$ (for SPP)
- exploit the explicit parallelism expressed in the PING

by minimizing a cost function by using a genetic approach [139] where each individual of the population represents a possible mapping/architecture item. Such a cost function is composed of several terms related to the following system-level metrics (partially described in Chapters 5, 6, and 8, and [150]) that are evaluated for each individual during the genetic evolution (a value of 0 indicates the best situation):

- *Affinity Index* (I_{A}): [0, 1]
  - Affinity between the features of the procedures and the processors on which they have been mapped to
- *Load Indexes* (I_{L}): [0, 1]
  - Balancing of the workload over the available GPP and DSP with respect to L_{MAX}

- *Communication Index* (I_C): [0, 1]
  - Exchanged data size between procedures mapped onto different BBs (i.e., exchanged data size between BBs)
- *Physical Cost Index* (I_{ϵ}): [0, 1]
  - Cost of the solution
- *Size Indexes* (I_{KB}, I_{Geq}): [0, 1]
  - Memory (for GPP/DSP) or resources (for SPP) utilization
- *Concurrency (Parallelism) Index* (I_P): [0, 1]
  - Exploitation of the potential concurrency expressed in the CSP: a value of 0 (i.e., the best) indicates that all the potential concurrent PING procedures are mapped onto different processors while a value 1 (i.e., the worst) indicates that all the potential concurrent PING procedures are mapped onto the same processors (i.e., no concurrency is exploitable).

By combining the previous indexes, a linear cost function has been built to compare different mapping/architecture items:

$$\text{CF} = w_A \cdot I_A + w_L \cdot I_L + w_C \cdot I_C + w_{\epsilon} \cdot I_{\epsilon} + w_{\text{KB}} \cdot I_{\text{KB}} + w_{\text{Geq}} \cdot I_{\text{Geq}} + I_P \cdot w_P$$

where the weights w (belonging to the interval [0,1]) are used to identify individuals that better tradeoffs different indexes or to perform Pareto analysis.

The structure of the individual is represented by an entry for each PING procedure: each procedure is associated with a type of processor and an instance number. Figure 10.8 shows a possible individual instance with its corresponding mapping/architecture for the PING of Figure 10.7. In general, the mapping could be formally expressed as done in [160].

The initial population is randomly generated, while, during the evolution of the population, the algorithm performs the optimizations that minimize the cost function following the classical rules of genetic algorithms. In particular, each *crossover* operation generates two new individuals combining two existing ones as shown in Figure 7.5 while the *mutation* operation changes randomly type and/or instance number of a processor associated with a randomly selected procedure.

During the evolution, the individuals that score the worst values tend to be replaced by better ones. Several parameters in the algorithm (e.g., population size, number of generations, mutation probability, etc.) allow a wide exploration of the design space, with the goal to avoid local minima.

P1	P2	P3	P4	P1.1	P1.2	P3.1	P4.1
GPP	SPP	GPP	DSP	SPP	DSP	GPP	GPP
1	2	1	1	1	1	2	2

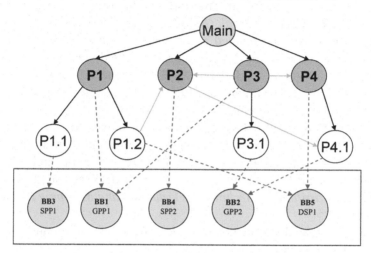

Figure 10.8 An individual and its corresponding architecture.

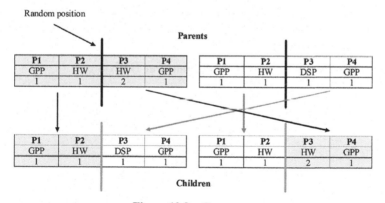

Figure 10.9 Crossover.

It is worth noting that, to avoid complexity explosion while coping with large system specifications, it is possible to exploit the clustering approach already described in Chapter 7 and [150] (currently not implemented in the PAM1 tool).

10.5.2 Second Phase

The output of the first phase is so related to the computation aspects of the architecture: number and type of BB/PUs in the architecture and the mapping of each PING procedure onto them. The starting point of the second phase is the so-called *BBs Interaction Graph (BING)*, i.e., an internal model used to represent the partial system obtained at the end of the first phase. It allows to shift from a procedure-oriented view to a BB-oriented view of the system while keeping the information related to the BBs that need to communicate. Such BBs are so connected with an edge that will be annotated depending on a proper function (just f in Figure 10.10) of the bandwidth required by each procedure mapped onto each BBs (e.g., the sum of max values to represent the worst-case scenario, the sum of averages value to represent an average situation, etc.). Figure 10.10 represents the BING related to the architecture of Figure 10.8 for the PING of Figure 10.7. Such a model is provided as input to the *PAM2* (i.e., *Partitioning, Architecture Definition and Mapping Phase 2*) tool (Figure 10.6).

So, starting from a BING, the goal is to determine the number and type of ILs between BBs in order to:

- minimize cost of the set of ILs
    - The max number of instances for each IL can be specified by the designer as an architectural constraints (to limit cost or to model an existing platform)
- keep the bandwidth of each IL under but near BW_{MAX}

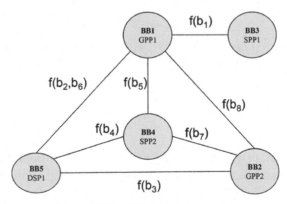

Figure 10.10   BBs interaction graph.

- keep the number of BB using each IL under but near N_{MAX} (and $>=$ N_{MIN})
- keeping feasibility by respecting CUs characterization

by minimizing a cost function by using a genetic approach where each individual of the population represents a possible interconnections/topology item. Such a cost function is composed of several terms related to the following system-level metrics (partially described in [166]) that are evaluated for each individual during the genetic evolution (a value of 0 indicates the best situation):

- *Saturation Index (I_{B})*: [0, 1]
    - Respect of the max bandwidth offered by the IL
- *Exploitation Index (I_{E})*: [0, 1]
    - Respect of the min/max number of BBs that can use an IL instance
- *Physical Cost Index ($I_{\text{€}}$)*: [0, 1]
    - Cost of the solution
- *Concurrent Communications Index (I_{CC})*: [0, 1]
    - Respect of the max number of concurrent communications allowed by the IL
- *Feasibility Index (I_{F})*: [0, 1]
    - A pair of CUs should be able to manage at least a common IL in order to allow the related BBs to directly communicate: such an index indicates how much of the actual communications are unfeasible (as described in the following)

By combining the previous indexes, a linear cost function has been built to compare different interconnections/topology items:

$$\mathrm{CF} = w_B \cdot I_B + w_E \cdot I_E + w_{\text{€}} \cdot I_{\text{€}} + w_{\mathrm{CC}} \cdot I_{\mathrm{CC}} + w_{\mathrm{F}} \cdot I_{\mathrm{F}}$$

where the weights w (belonging to the interval [0,1] except for w_F whose upper bound could be freely fixed by the designer to penalize unfeasible interconnections/topology item) are used to compare different individuals in order to identify the one that better tradeoffs different parameters or to perform Pareto analysis.

The structure of an individual of the population is represented by a triangular matrix with the number of rows and columns equal to the number

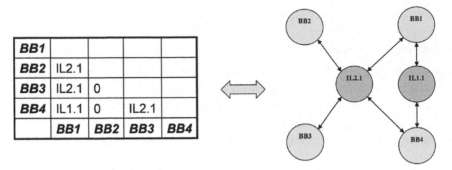

BB1				
BB2	IL2.1			
BB3	IL2.1	0		
BB4	IL1.1	0	IL2.1	
	BB1	*BB2*	*BB3*	*BB4*

Figure 10.11   An example of individual and corresponding architecture.

of BBs. Each couple of communicating BBs is associated with an instance of an IL.

For example, the individual of Figure 10.11 represents an architecture with 4 BBs that communicates by means of an instance of IL2 (i.e., *BB1* and *BB2*, *BB1* and *BB3*, and *BB3* and *BB4* all use *IL2.1*) while an instance of IL1 (i.e., *IL1.1*) connects *BB1* and *BB4*. The 0 values represent couples of BBs that do not need to communicate directly.

In order to perform the exploration of the design space, the initial population is randomly generated, while during the evolution of the population, the algorithm follows the classical rules of genetic algorithms. In particular, it is very important to analyze the adopted cross-over approach: starting from two individuals, it is possible to apply both a vertical (Figure 10.12 left) and an horizontal (Figure 10.12 right) cut in order to preserve respectively columns and rows. In this way, each couple involved in the cross-over gives rise to 4 new individuals, thus spreading and accelerating the design space exploration.

However, such an approach could give rise to unfeasible solutions so the feasibility of the children has to be checked and properly took into account: unfeasible solutions will get a higher cost function and so they will be probably removed from the population. However, there is always the possibility of a mutation process giving rise to better individuals that otherwise could be difficult to obtain (i.e., avoiding local minimum). In order to check such a feasibility, it is needed a CU characterization with respect to the BBs belonging to the BING. This is obtained by defining a proper *CU Characterization Matrix* (CUCM) that explicitly indicates the ILs that each BB is able to manage. A generic example of such matrix is shown in Figure 10.13.

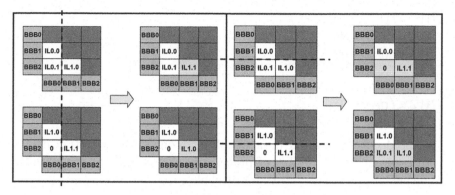

Figure 10.12 Cross-over.

	IL1	IL2	IL3	IL4	IL5
BB1	X	X	X		X
BB2	X	X	X	X	X
BB3	X		X		X
BB4	X		X	X	
BB5		X	X	X	X

Figure 10.13 CU characterization matrix.

10.5.3 Illustrative Example

In order to support the presented approach by checking the consistency of the results, two C++ tools (i.e., PAM1 and PAM2) have been developed and several tests have been performed. In order to clarify the main features of the whole approach, a simple example is reported in the following. The focus is mainly on Phase 2, since a lot of details on Phase 1 could be found in Chapter 7 and [150].

Let the CSP specification be the one shown in Figure 10.3 and the related annotated PING the one shown in Figure 10.7. Let the annotated values of a_i, l_i, s_i, and b_i be such that the *PAM1* tool would provide as output of the first phase the mapping/architecture shown in Figure 10.8. The *BING* to consider is then the one shown in Figure 10.10. Now, let us fix some fictitious values in order to make the tool *PAM2* able to apply the second phase and to provide some results.

First, it is needed to assign some values to the bandwidths (e.g., MB/s) in the *BING*:

- $BW_1 = f(b_8) = 250$
- $BW_2 = f(b_1) = 200$
- $BW_3 = f(b_5) = 200$
- $BW_4 = f(b_7) = 150$
- $BW_5 = f(b_2, b_6) = 250$
- $BW_6 = f(b_3) = 50$
- $BW_7 = f(b_4) = 50$

Then, it is needed to characterize the available interconnections links (i.e., max bandwidth, min and max number of processors, max number of concurrent communications, and relative cost), for example:

- $IL_1 = (300, 2, 2, 1, 1)$
- $IL_2 = (400, 2, 2, 2, 2)$
- $IL_3 = (250, 3, 8, 1, 3)$
- $IL_4 = (400, 3, 6, 1, 4)$
- $IL_5 = (200, 4, 80, 4, 5)$

Actually, from an implementation point of view (i.e., a lower level of abstraction), the different ILs could be representative of point-to-point communications (*IL0/IL1*, e.g., *GPIO or UART based*), of a standard bus (*IL2/IL3*, e.g., I^2C or *SPI*), or of a mesh (*IL4*, e.g., *crossbar/omega switch* or *network-on-chip*). Next, it is necessary to associate such ILs with the CUs that are able to manage them: for example, let us keep valid the matrix of Figure 10.13. Hence, the second phase of the system design exploration is performed introducing, in the cost function described previously, the following trade-off weights (a Pareto analysis is also possible):

$$W_B = W_E = W_{CC} = 0.2, W_{\mathbb{C}} = 0.4, W_F = 100$$

and imposing, for example, the following architectural constraints: max number of instances for each kind of *IL* equal to 4.

Starting from a population of 1000 individuals (with a bound of 10000), an execution of the *GA* for 10000 generations gives rise to the following best result:

$$I_B = 0.42, I_E = 0.1, I_{\mathbb{C}} = 0.14, I_{CC} = 0.1, I_F = 0, \quad CF = 0.18$$

associated with the individual/architecture shown in Figure 10.14.

Such an architecture reflects the characterizations of ILs (i.e., bandwidth, number of processors, etc.) and CUs (i.e., characterization matrix), and represents a sub-optimal optimization for the adopted cost function.

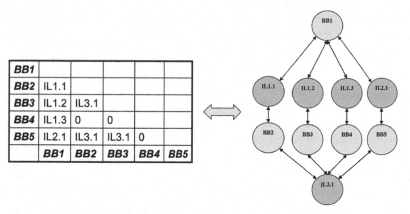

BB1					
BB2	IL1.1				
BB3	IL1.2	IL3.1			
BB4	IL1.3	0	0		
BB5	IL2.1	IL3.1	IL3.1	0	
	BB1	**BB2**	**BB3**	**BB4**	**BB5**

Figure 10.14 Best individual and corresponding architecture.

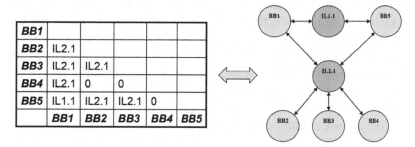

BB1					
BB2	IL2.1				
BB3	IL2.1	IL2.1			
BB4	IL2.1	0	0		
BB5	IL1.1	IL2.1	IL2.1	0	
	BB1	**BB2**	**BB3**	**BB4**	**BB5**

Figure 10.15 Best individual 2 and the corresponding architecture.

Now, with the goal to check the behavior of the algorithm when changing some parameters, let us modify the IL characterization introducing a new high-performance IL2 link at the same price:

- $IL_2 = (1000, 2, 8, 4, 2)$

Starting again from a population of 1000 individuals (with a bound of 10000), an execution of the GA for 10000 generations gives rise to the following best result:

$$I_B = 0.28, I_E = 0.19, I_\mathcal{E} = 0.07, I_{CC} = 0.11, I_F = 0, CF = 0.185$$

associated with the individual/architecture shown in Figure 10.15.

Such an architecture meets again all the constraints while exploiting the new link features to suggest a cheapest architecture.

To perform the presented design space explorations, the tool, executed on a Pentium M (1.2 GHz, 1 GB RAM), has requested an execution time less than 10 minutes.

10.6 Conclusion

This Chapter has coped with the problem of hw/sw co-design of dedicated digital systems based on heterogeneous multi-processor architectures. When such systems are implemented by means of a meaningful (i.e., hundreds) number of *processor-cores* in a single ICs, they are also called *Dedicated Heterogeneous Many-Core Systems*. In particular, the Chapter has proposed a system-level design space exploration approach able to suggest to the designer an HW/SW partitioning of the specification and a mapping of the partitioned entities onto an automatically selected heterogeneous multi-processor architecture, being aware not only of the computational issues but also to take into consideration the communication ones. The approach has been partially validated by means of proper tools by checking the consistency of the adopted metrics and heuristics. The obtained experimental results are encouraging and justify further research efforts in this direction. In fact, the current work is oriented on both the two phases of the described approach. For the first phase, since the *Affinity* metric (presented for the very first time in Chapter 5) has been exploited and improved by several works in the recent years (e.g., [163–165]), the goal is to integrate such improvements in the current methodology and also to introduce novel ones, e.g., to enrich the set of supported PUs (e.g., *Graphical Processing Unit*). For the second phase, the goal is to enhance the set of parameters and features used to characterize different ILs (e.g., latency, contention to access the IL, etc.) in order to be able to model advanced network aspects (e.g., multi-hop) and fully consider also *Network-on-Chip* architecture. Finally, with respect to the whole co-design methodology, a full validation with respect to real-world data coming from lower levels of abstraction will be of very critical importance.

11

SystemC-Based ESL Design Space Exploration for Dedicated Heterogeneous Multi-Processor Systems

11.1 Introduction

Systems based on heterogeneous multi-processor architectures (*Heterogeneous Multi-Processor Systems*, HMPS) have been recently exploited for a wide range of application domains, especially in the *System-on-Chip* (SoC) form factor (e.g., [143, 146, 178, 179, 180]). Such systems can include several processors, memories, and a set of interconnection links between them. By definition, the set of processors in the same architecture is heterogeneous. This implies that it is possible to exploit, at the same time, one or more processors belonging to different processing classes:

- *General-Purpose* Processors (GPP): e.g., *ARM, MIPS, MicroBlaze, NiosII*, Leon3, etc.
- *Domain-Oriented* Processors: e.g., *Digital Signal Processor* (DSP), *Graphical Processing Unit* (GPU), *Network Processor* (NP), etc.
- *Specific-Purpose* Processors (SPP): e.g., *AES encoder/decoder, JPEG encoder/decoder, UART/SPI/I2C controllers*; in general, every ad-hoc developed digital HW component.

Finally, such processors can be adopted in the form of *soft, hard,* or *fuse (i.e., hardwired) IP cores* or as *discrete IC* mainly depending on the final system form factor (i.e., *on-chip, on-FPGA, on-board*) and scope (final product or platform).

As stated in the title, this work focuses also on dedicated systems. In the scope of this Chapter, a *Dedicated System* (DS) is a digital electronic system with an application-specific HW/SW architecture. It is specifically designed in order to satisfy *a priori known* application requirements (both functional and not functional). A DS could be then *embedded* in a more complex system

199

and/or it could be subjected to *hard/soft real-time* constraints. When DS are also HMPS, they are called *Dedicated Heterogeneous Multi-Processor Systems* (D-HMPS).

Apart from possible differences on terminology and composition, for this kind of systems one consideration is always true: they are so complex that the adopted *HW/SW Co-Design Methodology* plays a major role in determining the success of a product. Moreover, in order to cope with such a complexity, the selected methodology should allow the designer to start working at the so-called *Electronic System-Level* (ESL) of abstraction. This means to be able to start the design activities from an executable model of the system behavior based on a given *Model of Computation* (MoC) that would be unifying for HW and SW, and that could be described by means of a proper specification/modeling language. In fact, as described also several in surveys (a recent one could be found in [181]), in the past years, a remarkable number of research works have focused on the (electronic) system-level HW/SW co-design of (D)HMPS (e.g., [150–156, 182]). In such works, the most critical issues are always related to the *System Specification* and *Design Space Exploration* steps. In the first one, the designer models the behavior of the desired system (specifying also possible non functional constraints), the available basic HW components, and the target HW architecture. The second step is then related to the approach, automated or not, used to find the best HW/SW partitioning and mapping for the final system implementation. The main differences between the various approaches are related to the different amount of information and actions that are requested to the designer and that are so heavily influenced by his experience. In particular, a lot of approaches (especially those based on the on the *Y-Chart* principle [188]) explicitly require as an input the HW architecture to be considered for mapping purposes. So, up to now, there are very few system-level HW/SW co-design flows that try to fully addresses the problem of both *"automatically suggest an HW/SW partitioning of the system specification"* and *"map the partitioned entities onto an automatically defined heterogeneous multi-processor architecture"*. Two of the works most similar to the proposed one are [155] and [182]. The first one presents a SystemC-based Co-Design flow that tries to automate the process of prototypes generation. The approach exploits a commercial synthesis tool to generate automatically hardware accelerators starting from a SystemC behavioral model. However, the designer has to manually describe an *Architecture Template* that represents the architectural infrastructure to be used during the DSE. The second one is still more interesting. In fact, it presents a very flexible and extensible system-level MP-SoC design space exploration infrastructure that is also able to automatically generate hardware architectures. By contrast, the approach presented in the proposed work is based on a more integrated and customized environment able to fully exploit from the very beginning, since they are evaluated before starting the DSE step, some metrics, and estimations, mainly related to performances, to perform in a more effective way the DSE. Moreover, the proposed approach is also able to explicitly consider also SPP; a processing class that is not clearly managed in [182]. Finally, to take a look also to a representative SystemC-based commercial product, it is worth citing *Intel CoFluent* [187] as a promising (electronic) system-level modeling and simulation environment. Other than the model of the system behavior, it still requires explicit and manual modeling of both the hardware architecture and the mapping but, thanks to its *Eclipse* based architecture, it is possible to think about some future plug-in extensions oriented to support the designer also in such activities.

According to this scenario, this Chapter presents the first meaningful attempt to derive, from the methodology described in the research path delineated in Part 1 and Chapter 10, a prototypal *"SystemC-based Electronic System-Level HW/SW Co-Design Environment for Dedicated Heterogeneous Multi-Processor Systems"* to be used to fully exploit and validate the proposed approach. In particular, the preliminary results described in [183] are better described and further refined, to overcome previous limitations, while also providing the relevant methodological background.

11.2 Reference ESL HW/SW Co-Design Flow

The reference ESL HW/SW co-design flow is shown in Figure 11.1 (it is a slight extension/modification of the ones described along Part 1 and in Chapter 10): it reports the main co-design steps and the needed information that are briefly described in the following paragraphs.

11.2.1 System Behavior Model

The entry point is a model of the system behavior (SBM) based on the *Communicating Sequential Processes* (CSP) model of computation [37, 159]. SBM represents the functional requirements while non-functional ones are currently related only to a *Time-to-completion* constraint (TTC) and the following architectural ones:

- a fixed set of available processors and interconnection links types;

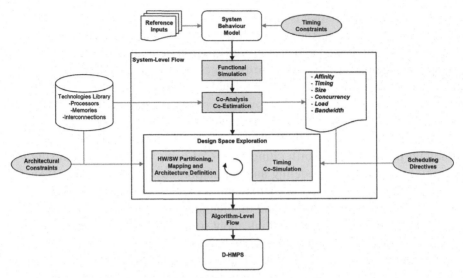

Figure 11.1 The reference co-design flow.

- min and max number of available processors and interconnection links instances;
- available area for ASIC or an equivalent metric for FPGA;
- a reference template HW architecture mainly composed of a set of rules that the tool has to follow while automatically building the final HW architecture; and
- available scheduling policies.

In the proposed approach, SBM is captured by means of a *procedure-level* internal model called *Procedural INteraction Graph* (*PING*) [150] while each procedure could be then described, at *statement-level*, by using a proper specification/modeling language suitable to represent CSP features (e.g., *OCCAM* [12], *HandelC* [184], *ImpulseC* [172], *SystemC* [14]). PING is a formalism that provides information about the relationships among procedures and the exchanged data. By means of a PING, it is then possible to represent a CSP while taking into account also possible decomposition of each CSP process in classical functions/methods. In such a sense, a PING is able to explicitly show, with a procedure-level coarse-grain view, concurrency and communication/synchronization issues and it is so suitable as main input, when further annotated with several metrics and estimations, for the design space exploration step. Figure 11.2 graphically shows a CSP (with 6 processes and 6 channels) and a possible related PING where the processes have been arbitrarily decomposed in further elements (i.e., procedures).

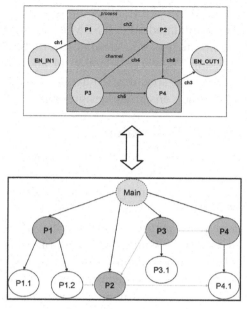

Figure 11.2 CSP and a related PING.

11.2.2 Technologies Library

In order to list and describe the basic HW elements available to automatically build the final HW architecture, a proper *Technologies Library* (TL) provides a characterization of available processors, memories, and interconnection links. Such a library contains information like processing classes (i.e., actually only GPP, DSP, and SPP), operating frequencies, maximum load (for GPP and DSP classes), capacity (for SPP and memories), max bandwidth (for interconnection links), relative cost (considering both the cost related to obtain a component and the effort needed to use it), and so on. Such information is then considered during the different steps of the co-design flow.

11.2.3 Functional Simulation

The first step of the proposed co-design flow is the *Functional Simulation* where SBM is simulated to check its correctness with respect to some *Reference Inputs*. Such data sets are of critical importance since they have to be as much as possible representative of the actual operating conditions of the system. Such a simulation is normally very fast and it allows also to take into account timed inputs, i.e., there is a concept of simulated time, but it does not consider the time needed to execute the statements, related to both computation and communication operations, composing the processes, i.e., statements are executed in 0 simulated time. If SBM is not correct (i.e., wrong outputs or critical conditions such as e.g., deadlocks), it should be properly modified and simulated again. The early detection of anomalous behaviors allows the designer to correct the specification avoiding a late discovery of problems that could lead to time-consuming design loops.

11.2.4 Co-Analysis and Co-Estimation

This step aims at extracting as much as possible information about the system by analyzing the SBM while considering the provided TL. This step is composed of *Co-Analysis* (Chapter 5 and [150]) and *Co-Estimation* activities (Chapters 6, 8, and [167–169]).

Co-Analysis provides a set of metrics expressing the *Affinity* of each CSP process toward the given set of processing classes, and some information about *Concurrency*. In particular, the latter identifies the set of CSP processes and communications that can be potentially executed concurrently.

Co-Estimation provides a set of estimations about *Timing*, *Size*, *Load*, and *Bandwidth*. *Timing* is related to the estimation of the number of clock cycles needed, by each processor in TL, to execute each single statement composing the processes in SBM. *Size* represents the number of ROM/RAM bytes needed for SW implementations and *equivalent gates* (or similar metrics for FPGA) for HW ones. Finally, by exploiting Timing data and considering the TTC constraint, it is also possible to estimate the *Load* associated with the execution of the SBM processes when mapped on a single instance of each processor in TL, and the *Bandwidth* needed to the different processes to communicate while fulfilling the TTC constraint. The extraction of these data from the SBM is an important step that allows, during the following design space exploration, the identification of the number and processing class of processors needed to satisfy the TTC constraint.

11.2.5 Design Space Exploration

Finally, the reference co-design flow reaches the *Design Space Exploration* (DSE) step [150, 166] that is constituted of two iterative activities: "*HW/SW Partitioning, Mapping and Architecture Definition*" and "*Timing Co-Simulation*". All the data (i.e., metrics and estimations) extracted in the previous steps are then used to drive the DSE, together with additional information/constraints provided by the designer: available *Scheduling Directives* (i.e., available scheduling policies and possible priorities among processes) and possible *Architectural Constraints* (i.e., max number of instances for each available processor and interconnection link). The *HW/SW Partitioning, Mapping and Architecture Definition* activity is based on a genetic algorithm that allows to explore the design space looking for feasible mapping/architecture items suitable to satisfy imposed constraints. Then, the *Timing Co-Simulation* activity considers suggested mapping/architecture items to actually check for TTC constraint satisfaction. If the suggested mapping/architecture item does not meet such a constraint, the designer should perform again the design space exploration by changing some exploration parameters, by modifying the starting SBM, by enriching the TL with new elements, or by relaxing some constraints.

11.2.6 Algorithm-Level Flow

When the mapping/architecture item proposed by the DSE step is satisfactory, it is possible to implement the system. For this, the SW-mapped processes are typically transformed in C code, with the support of a possible embedded and/or real-time OS, while the HW-mapped ones are transformed in synthesizable HDL code or implemented by means of existing COTS component depending on the final system form factor. It is worth noting that such transformations will be done automatically or manually depending on the language and the coding style adopted to describe the SBM. This step is fully based on existing commercial algorithm-level methodologies and tools that are out of the scope of this work.

11.2.7 Reference Template HW Architecture

The reference template HW architecture selected for the proposed methodology is a heterogeneous multiprocessor one with distributed memory [166]. So, the final system will be composed of proper interconnections of some instances of different basic elements. The basic elements, called *Basic Block* (BB), represent the minimal computation, storage, and communication units in the system. They could be different in their internal components giving so rise to possible heterogeneous multiprocessor systems. In particular, each BB is composed of three main elements: a *Processing Unit* (PU), a *Local Memory* (LM), and a *Communication Unit* (CU). Finally, a PU could actually belong only to one of the three different processing classes (i.e., GPP, DSP, and SPP). All the elements in a BB are characterized by means of the data available in the TL. So, given some instances of BBs and interconnecting them by means of some instances of interconnections links, the tool is able to automatically define a feasible dedicated heterogeneous multi-processor architecture on which the system functionalities can automatically be mapped to. In the proposed approach, such an architecture is then represented, for design space exploration purposes, by means of an internal model based on the so-called *Architecture Graph* [160].

11.3 SystemC-Based ESL HW/SW Co-Design Environment

As said in the introduction, this chapter proposes the first meaningful attempt to derive, from the methodology briefly recalled (described with more detail in Part 1 and Chapter 10), a fully working *"SystemC-based Electronic System-Level HW/SW Co-Design Environment for Dedicated Heterogeneous Multi-Processor Systems"*. So, the following paragraphs describe the main customizations that have been performed to adapt to *SystemC* the reference co-design flow.

11.3.1 System Behavior Model

Since SBM is based on CSP, the *SystemC* library has been extended to properly model CSP processes and, in particular, CSP channels (taking inspiration by the works presented in [173–175]). In fact, while CSP processes are modeled by exploiting basic SC_THREAD, CSP channels have been modeled by introducing a proper SC_CSP_CHANNEL in the *SystemC* library. In particular, a CSP process is an SC_THREAD presenting an infinite loop behavior. It is able to directly access only to its local variables and so it communicates with other CSP processes only by means of CSP channels. Moreover, in the considered SC_THREAD, only basic *C/C++* statements and *SystemC* data types are allowed while avoiding a full OOP approach since it could introduce critical issues for estimation and HW synthesis activities. Finally, the CSP channel (i.e., SC_CSP_CHANNEL) has been obtained by modifying the SC_FIFO while providing an interface that offers blocking *write()* and *read()* methods (Figure 11.3 shows such an interface and, as an example, the implementation of the blocking *read*). Main modifications with respect to the SC_FIFO are related to the introduction of a full-handshake protocol to allow synchronous data exchange, as expected for a CSP channel.

It is worth noting that, in order to allow an SC_THREAD to check for data from more than a channel at the same time (how it is possible, e.g., with the ALT statement of the OCCAM language [12], also based on CSP), two more methods, i.e., *read_test()* and *write_test()*, have been

```
/***************************************************/        // Blocking read
sc_csp_ifs.h -- The sc_csp_channel<T> interface classes.
/***************************************************/        template <class T>
#ifndef SC_CSPCHANNEL_IFS_H                                  inline
#define SC_CSPCHANNEL_IFS_H                                  void
#include "sysc/communication/sc_interface.h"                sc_csp_channel<T>::read( T& val_ )
                                                             {
namespace sc_core                                                  if(ready_to_write==true)
{                                                                  {
  template <class T>                                                   ready_to_read=true;
  class sc_csp_channel_in_if:virtual public sc_interface              ready_to_read_event.notify(SC_ZERO_TIME);
  {                                                                    sc_core::wait(ready_to_write_event);
      public:
          // read                                                      val_=csp_buf;
          virtual void read( T& ) = 0;
          virtual T read() = 0;                                        ready_to_read=false;
      protected:                                                       ready_to_read_event.notify(SC_ZERO_TIME);
          // constructor                                           }
          sc_csp_channel_in_if() {}                             else
  };                                                             {
                                                                     ready_to_read=true;
  template <class T>                                                 ready_to_read_event.notify(SC_ZERO_TIME);
  class sc_csp_channel_out_if:virtual public sc_interface            sc_core::wait(ready_to_write_event);
  {
      public:                                                        val_=csp_buf;
          // blocking write
          virtual void write( const T& ) = 0;                        ready_to_read=false;
      protected:                                                     ready_to_read_event.notify(SC_ZERO_TIME);
          // constructor                                             sc_core::wait(ready_to_write_event);
          sc_csp_channel_out_if(){}                              }
  };                                                           }
} //end of namespace
#endif
```

Figure 11.3 SC_CSP_CHANNEL interface and *read()* implementation.

added to SC_CSP_CHANNEL as a base to build an ALT-like CSP construct. All these elements allow an effective electronic system-level modeling of D-HMPS behavior. In particular, the whole system behavior is enclosed into a single SC_MODULE containing all the CSP processes and channels. Other SC_MODULE and SC_CSP_CHANNEL are then used to model the *Test-Bench* and connected to the system by means of proper SC_PORT. A schematic example is shown in Figure 11.4.

11.3.2 Functional Simulation

Since the *SystemC* model representing SBM is executable by construction, this step is straightforward and it is directly based on the simulation kernel provided by the standard *SystemC* library (commercial simulators can be used as well).

11.3.3 Co-Analysis and Co-Estimation

This step is mainly based on techniques and tools already introduced in Part 1 and Chapter 10. One of the main innovation is related to the use of a *SystemC* timing co-simulation, as described in the next section, to perform load estimation (as described in Chapter 8) and to automatically identify the set of CSP processes and communications that can be potentially executed concurrently.

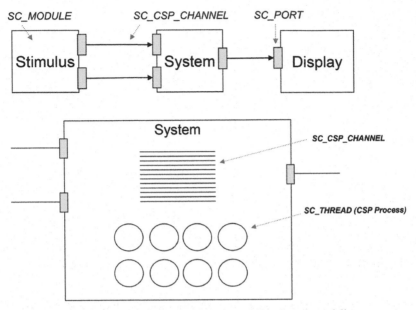

Figure 11.4   An example of System and Test-Bench modeling.

11.3.4 Design Space Exploration

Since such a step is the main focus of this chapter, it is described with more detail in the next paragraph.

11.4 SystemC-Based ESL Design Space Exploration

As described before, this step is composed of two iterative activities: *"HW/SW Partitioning, Mapping and Architecture Definition"* and *"Timing Co-Simulation"*. The final goal is the automatic identification of:

- a HW/SW partitioning of the CSP processes;
- a heterogeneous multi-processor architecture composed of several connected BB picked up from the TL and able to satisfy the architectural constraints;
- a mapping of the partitioned CSP processes onto the identified BB able to satisfy the TTC constraint.

It is worth noting that the target HW architectures considered for the co-design environment developed up to now, are currently limited to be *heterogeneous mono-core multi-processor* ones, where each processor has its own local memory (i.e., the system has a distributed memory architecture). Moreover, processors are currently able to communicate only by means of a shared bus (i.e., buses are the only interconnection links actually contained in the TL). In other words, the communication architecture is fixed and so the second phase of the DSE approach (Figure 10.6) introduced in [166] is actually not fully supported.

11.4.1 HW/SW Partitioning, Mapping, and Architecture Definition (1st Phase)

The first phase is related to the mapping of SBM (i.e., a CSP specification written in SystemC and represented by means of a PING) onto a dedicated architecture by following the approach described in Part 1 and [150]. The internal-model representing the specification, annotated by means of the *Co-Analysis and Co-Estimation* step, is provided as input to the *PAM1* (i.e., *Partitioning, Architecture Definition, and Mapping Phase 1*) tool. So, starting from an annotated PING, the first phase goal is to determine the number and type of BB/PUs and a mapping of PING procedures onto them while minimizing a cost function using a genetic approach [139], where each individual of the population represents a possible mapping/architecture item (each procedure is associated with a type of processor and an instance number as shown in Figure 10.8). The considered cost function is composed of several terms related to system-level metrics that are evaluated for each individual during the genetic evolution.

In fact, the initial population is randomly generated, while during the evolution of the population the algorithm performs the optimizations that minimize the cost function following the classical rules of genetic algorithms (i.e., *crossover* and *mutation*). In particular, each crossover operation generates two new individuals combining two existing ones as shown in Figure 10.9 while the mutation operation randomly changes type and/or instance number of a processor associated with a randomly selected procedure. During the evolution, the individuals that score the worst values tend to be replaced by better ones. Several parameters in the algorithm (e.g., population size, number of generations, mutation probability, etc.) allow a wide exploration of the design space, with the goal to avoid local minima. It is worth noting that, to avoid complexity explosion while coping with large system specifications, it is possible to exploit the clustering approach already described in Chapter 7 (currently not implemented in the tool described below). The output of the first phase is so related to the computation aspects of the architecture: the number and type of BB/PUs in the architecture and the mapping of each PING procedure onto them.

To fully support the *1st Phase* in the context of the proposed SystemC-based co-design environment, PAM1 tool has been re-designed while keeping in mind two main goals:

- easy integration in the whole co-design flow;
- highly re-use opportunity for the development of PAM2 tool (2nd *Phase*).

The main design issues are briefly described below.

11.4.1.1 Inputs modeling

Main inputs to the PAM1 tool (i.e., annotated PING and parameters for the genetic evolution) are provided by means of two XML files. The UML model related to the annotated PING XML schema is shown in Figure 11.5.

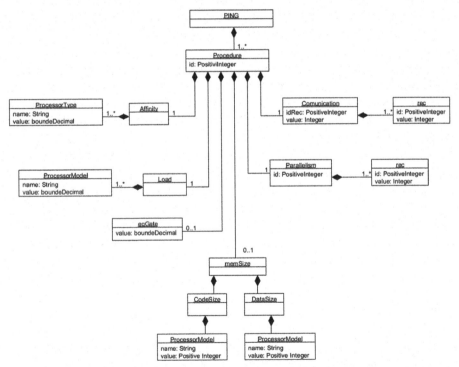

Figure 11.5   UML model related to the annotated PING XML schema.

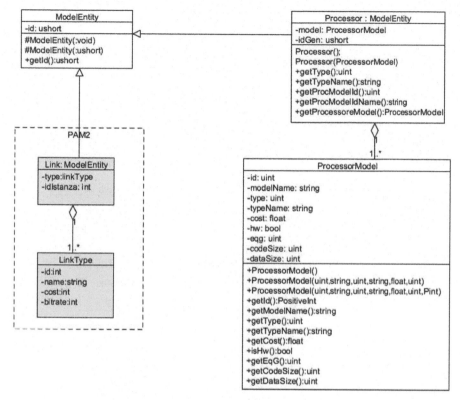

Figure 11.6 Technologies library modeling.

11.4.1.2 Technologies library modeling

Other than the annotated PING, there is the need to model also available processors and interconnection links (i.e., the TL). The adopted approach, shown in Figure 11.6, takes also into account requirements for PAM2 tool (*2nd Phase*) by exploiting abstract classes and polymorphism.

11.4.1.3 PAM specification modeling

The set of inputs (XML files and TL) becomes the starting point to build the population for the genetic algorithm (i.e., the specification for the particular DSE step). With the approach shown in Figure 11.7, the genetic algorithm engine is able to evolve the population by means of the specification classes independently of the specific phase (1st or 2nd).

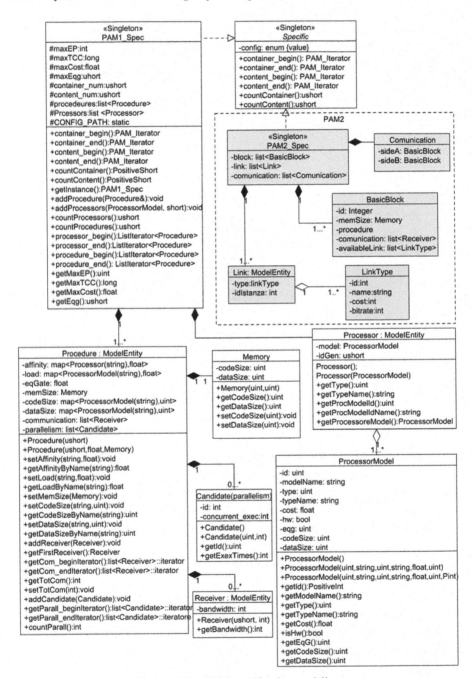

Figure 11.7   PAM specification modeling.

11.4.1.4 Optimization engine, individuals, and allocation modeling

Since the genetic approach is the main common element in both PAM1 and PAM2 tools, also if they rely on different individuals to perform different cost functions optimization, the design has been oriented to allow a meaningful reuse. The main issue is that in both cases there is the need to find a (sub)optimal allocation between two sets of objects: procedures to processors in the first case and the derived communications between processors to interconnection links in the second one. So, the individuals and the allocation modeling have been performed to exploit this common point of view as shown in Figure 11.8.

In fact, the optimization engine works with a set of generic individuals and so it can be simply reused by specializing the cost function evaluator as shown in Figure 11.9. This approach makes easy to change the algorithm for the optimization as well as to implement various cost function evaluators to explore the design space in a different way. It is worth noting that the use of a custom implementation (instead of using existing general libraries such as *GAlib* [140]) has been chosen to better exploit synergies with the proposed environment.

11.4.2 Timing Co-Simulation

As preliminarily introduced in [183], the *Timing Co-Simulation* is performed by means of an innovative *HW/SW Timing Co-Simulator*. In fact, it has been specifically built on the base of standard *SystemC* library by introducing some additional classes (i.e., *SystemManager*, *SimulationManager*, and *SchedulingManager*, as represented in Figure 11.10). Thanks to them, the current co-simulator is able to take into account a heterogeneous multiprocessor architecture, a processes-to-processors-to-links mapping, and all the relevant information previously collected, to check if a given TTC constraint is going to be satisfied. Additionally, the designer can select a scheduling policy to be used for processes implemented in SW and allocated on the same processor. With more details, *SystemManager* allows to take into account all the details about the system to be simulated (i.e., number and type of processors, number and type of interconnection links, *Affinity* and *Timing* data needed for simulation, mapping, etc.), while *SimulationManager* allows to configure the simulation type (i.e., functional or timing) and verbosity, and defines a set of *macro* used to instrument the simulated *SystemC* code. Finally, *SchedulingManager* allows taking into account the effects of the possible scheduling activities.

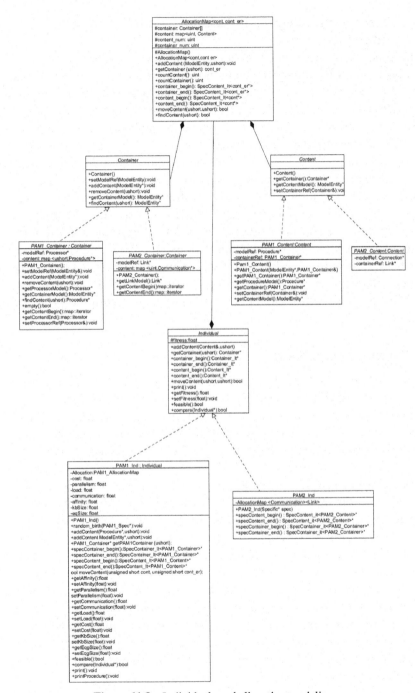

Figure 11.8   Individuals and allocation modeling.

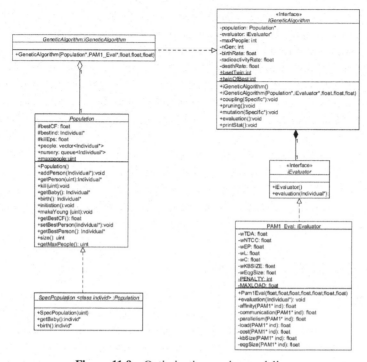

Figure 11.9 Optimization engine modeling.

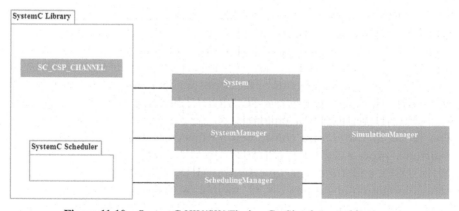

Figure 11.10 SystemC HW/SW Timing Co-Simulator architecture.

With respect to the scheduling activities modeling, the work has been inspired by the ones described in Chapter 8, and [161, 162]. As already said, the designer can select a scheduling policy for processes implemented in SW and allocated on the same processor. This scheduler acts as a user-level one with respect to the *SystemC* simulation kernel. It can be considered a *statement-level scheduler* since the approach adopted to model the scheduler behavior is based on the interactions with the *SchedulingManager* for the simulation of each statements belonging to a process (based on the exchange of *release/notify* events as done in [162]). With respect to the approach adopted in Chapter 8, this one enables a refined scheduler modeling since it allows to adopt any scheduling policies (they should be just added to the *Scheduling-Manager* itself) and also different policies on different processors. To allow all this, proper *awk/sed* scripts perform semi-automatic instrumentations of the SystemC code representing the SBM by inserting several macros used to take into account waiting times dependent on the mapping and the interactions with the scheduler (e.g., let you note macros I () and P () in the right side of Figure 11.12).

11.5 FIR-FIR-GCD Case Study

In order to show the main features of the proposed *SystemC-based Electronic System-Level HW/SW Co-Design Environment*, a reference case study (the "*FIR-FIR-GCD*", already used in [183]) is reported below. It is worth noting that the considered example does not perform a meaningful computation but it is just used as a simple example to easily understand the potential of the proposed approach.

Let the SBM be, represented by the CSP shown in Figure 11.11, composed of eight CSP processes and twelve CSP channels. Two more processes and three more channels are then used to describe and connect the test-bench. Figure 11.4 provides a schematic view of such a system while Figures 11.12 and 11.13 show some relevant parts of the correspondent *SystemC* description: the *main()* function, the *mainsystem* SC_MODULE, and a CSP process (i.e., an SC_THREAD that uses CSP channels).

The *Technologies Library* considered for this case study is composed of three different processors: *Intel MPU8051* (16 MHz, GPP), *Microchip PIC24* (32 MHz, DSP), and *Xilinx Spartan3AN* (50 MHz, SPP). TL contains all the relevant information about processors, memories (local to processors), and interconnections (in this case study only shared buses) needed to perform the DSE. The selected form factor for the final system implementation is the *on-board* one and so each one of the available processors will be adopted as discrete IC.

At the first step of the co-design methodology, *Functional Simulation* allows to check SBM correctness by analyzing outputs obtained by given *Reference Inputs*. Figure 11.14 shows the outputs related to input stimulus

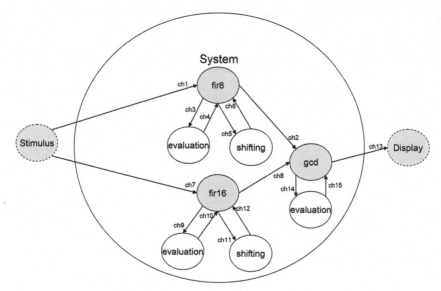

Figure 11.11   CSP representing the SBM.

```
SC_MODULE(mainsystem)
{
    // Ports for testbench connections
    sc_port< sc_csp_channel_in_if< sc_uint<8> > >
        stim1_channel_port;
    sc_port< sc_csp_channel_in_if< sc_uint<8> > >
        stim2_channel_port;
    sc_port< sc_csp_channel_out_if< sc_uint<8> > >
        result_channel_port;

    // PROCESSES
    void fir8_main();
    void fir8_evaluation();
    void fir8_shifting();
    ...

    // CHANNELS
    // fir8
    sc_csp_channel< fir8e_parameters >
        *fir8e_parameters_channel;
    sc_csp_channel< fir8e_results >
        *fir8e_results_channel;
    ...

    SC_CTOR(mainsystem)
    {
        SC_THREAD(fir8_main);
        SC_THREAD(fir8_evaluation);
        SC_THREAD(fir8_shifting);
        ...
```

```
//f8s
void mainsystem::fir8_shifting()
{
    // datatype for channels
    fir8s_parameters fir8s_p;
    fir8s_results fir8s_r;
    // local variables
    sc_uint<8> sample_tmp;
    sc_uint<8> shift[8];

    while(1)
    {
        // read parameters from channel
        I(f8s) fir8s_p=fir8s_parameters_channel->read();

        // fill local variables
        sample_tmp=fir8s_p.sample8; j=0;
        for( unsigned j=0; j<TAP8; j++)
            shift[j]=fir8s_p.shift[j];

        // processing
        I(f8s)
        for(int i=TAP8-2; i>=0; i--)
        {I(f8s)
            I(f8s) shift[i+1] = shift[i];
        }
        I(f8s) shift[0]=sample_tmp;

        // fill datatype
        for( unsigned j=0; j<TAP8; j++)
            fir8s_p.shift[j]=shift[j];

        // send results by channel
        I(f8s) fir8s_results_channel->write(fir8s_r);

        P(f8s)
    }
}
```

Figure 11.12   Sketches of SystemC descriptions of the main SC_MODULE and a CSP process.

provided with a *1000 ns* delay. It is worth noting as the output related to the *Display* process is obtained as soon as an input pair is available since the simulation is functional.

```
// Main function

int sc_main (int, char *[])
{
    ...

// Manager

    // System
    pSystemManager = new SystemManager("mysystemmanager");
    // Simulation
    pSimulationManager = new SimulationManager("mysimulationmanager");
    // Scheduling
    pSchedulingManager = new SchedulingManager("myschedulingmanager");

// Testbech and System

    // Channels for the connection to the main system
    sc_csp_channel< sc_uint<8> >    stim1_channel(8, s, f8m, COMMWIDTH1, TCOMM1, TACCCOMM1, SIMULATION_TYPE);
    sc_csp_channel< sc_uint<8> >    stim2_channel(8, s, f16m, COMMWIDTH1, TCOMM1, TACCCOMM1, SIMULATION_TYPE);
    sc_csp_channel< sc_uint<8> >    result_channel(8, gcdm, d, COMMWIDTH1, TCOMM1, TACCCOMM1, SIMULATION_TYPE);

    // Instantiation and connection of testbench and system
    stim_gen mystimgen("mystimgen");
    mystimgen.stim1_channel_port(stim1_channel);
    mystimgen.stim2_channel_port(stim2_channel);

    mainsystem mysystem("mysystem");
    mysystem.stim1_channel_port(stim1_channel);
    mysystem.stim2_channel_port(stim2_channel);
    mysystem.result_channel_port(result_channel);

    display mydisplay("mydisplay");
    mydisplay.result_channel_port(result_channel);

// Simulation management
    // Start simulation
    sc_start();

    ...

    // Total simulated time
    end=sc_time_stamp();
    ...

}
```

Figure 11.13   Sketch of SystemC *main*().

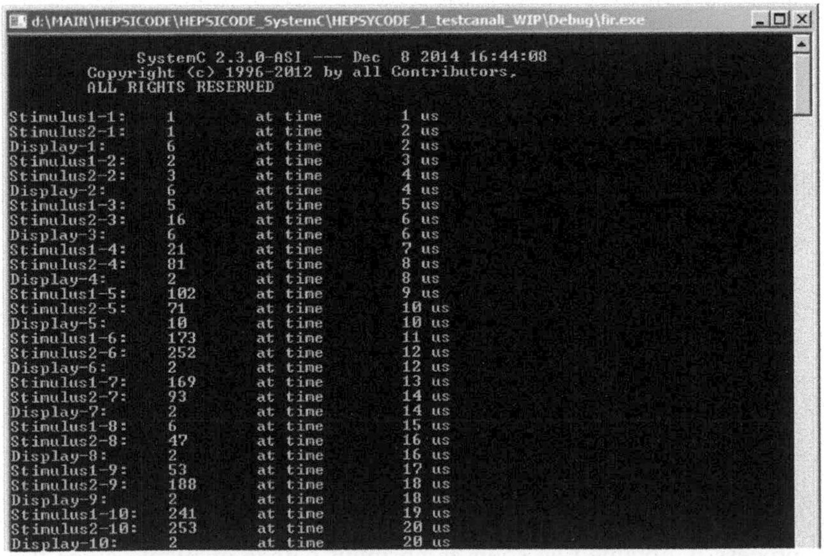

Figure 11.14   Outputs from functional validation.

Co-Analysis and *Co-Estimation* results activities are mainly the same performed in [183].

In particular, in this case, the Co-Analysis activity has been performed manually by the designer. In fact, based on his experience, he has provided values for the *Affinity* (Figure 11.15). On the other hand, potential *Concurrency* has been extracted automatically by means of proper co-simulations. From Figure 11.11, it is possible to see (by means of different colors) which processes are never run concurrently, e.g., *fir8-evaluation-shifting* processes cannot be concurrent (due to data dependency) and so on. Such information are so reported into the PING internally representing the SBM/CSP and the DSE step will take this into account in order to try exploiting only the real concurrency opportunities among CSP processes and CSP channels.

With respect to the *Co-Estimation* activity, it has been performed by means of some benchmarks (e.g., by using some reference code from [185] and some reference VHDL designs), some tools/boards (i.e., an *Instruction Set Simulator* for MPU8051, the *Evidence FLEX board* for PIC24, and the *Xilinx Spartan3AN DevKit* for HW implementation), and partially exploiting techniques already presented in [167–169]). The results about *Timing* are several *min-max* pairs (one for each processor) related to the number of clock cycles needed to execute the *SystemC* statements composing each CSP process. The actual values to be used during timing co-simulation are then dependent on mapping and *Affinity*. In fact, the *Affinity* values are used to perform an interpolation starting from *min-max* pairs. Similarly, *Size* data are *min-max* pairs related to the number of bytes needed for code/data and, since this case study refers to a FPGA as SPP, to the number of *Slices/LUT*. Finally, *Load* estimations for MPU8051 and PIC24 are performed by means of timing co-simulations (with *statement-level scheduling* and round robin policy) with respect to a set of TTC constraints (as described below). The final results are a set of estimated loads (for 8051 and PIC24) for each CSP process.

f8m	=	{0.9, 0.7, 0.5}
f8e	=	{0.5, 0.7, 0.5}
f8s	=	{0.5, 0.8, 0.9}
f16m	=	{0.9, 0.7, 0.5}
f16e	=	{0.5, 0.7, 0.7}
f16s	=	{0.5, 0.8, 0.9}
gcdm	=	{0.9, 0.7, 0.5}
gcde	=	{0.5, 0.7, 0.7}

Figure 11.15 Affinity with respect to GPP, DSP, and SPP.

As highlighted before, in this case study, the communication infrastructure has been fixed (i.e., processors with distributed memory and shared bus). So, the timing co-simulator directly takes into account the characterization data, related to the selected shared bus, provided in TL. In this case, an I^2C bus with a bandwidth fixed to 400 Kbps (more than sufficient to satisfy the communication requirements of the proposed example) has been adapted, since it is a multi-master one and it is available for all the considered processors.

Once all the metrics and all the estimations needed for the DSE step are collected, the following additional constraints are imposed:

- *Timing Constraints*: once estimated a *Worst Case Time-To-Completion* (i.e., WCTTC = 5.4 ms, Figure 11.16) by means of a timing co-simulation performed allocating all the CSP processes on a single MPU8051 instance, the DSE step will suggest a set of architecture/mapping pairs able to provide TTCs equal, respectively, to: 0.75*WCTTC, 0.5*WCTTC, 0.35*WCTTC, 0.25*WCTTC, and 0.1*WCTTC.
- *Architectural Constraints*: the DSE step can use max 4 instances of MPU8051, max 2 instances of PIC24, and max 1 instance of Spartan3AN FPGA.
- *Scheduling Policy*: processes implemented in SW and allocated on the same processor are subjected to round-robin policy with 10% time

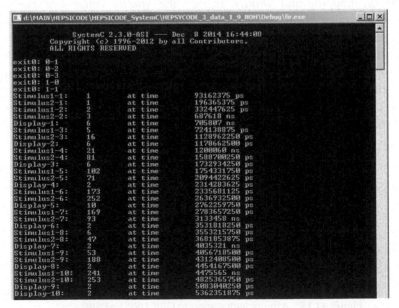

Figure 11.16   Worst-case time-to-completion.

overhead for context change (this will be simulated by means of a *statement-level scheduling*).

Given the previous set of TTC constraints, the DSE step provides the results shown in Figure 11.17. Starting from WCTTC (single MPU8051), when the requirement is 0.75* WCTTC, the DSE tool proposes a dual MPU8051 architecture allocating separately processes 0-1-2 and 3-4-5-6.7. The timing co-simulation estimates a TTC equal to 2.94 ms that well satisfies the 4.05 ms TTC constraint. Imposing 0.5* WCTTC, the DSE step proposes a triple MPU8051 architecture with the allocations 0-1, 3-4-5 and 2-6-7. However, this leads to an estimated time a bit greater than 2.7 ms. Performing a new DSE step with different parameters in the genetic algorithm and different weights in the adopted heuristic, the result is an architecture composed of a PIC24 with an MPU8051 that allows to satisfy the TTC constraint (estimated time is 2.43 ms) but at higher cost. Imposing 0.35*WCTTC, the DSE step finds another dual MPU8051-PIC24 architecture with an allocation that satisfies the new constraint. It is worth noting that such a mapping/architecture item is similar to the previous one (but with a different allocation) but, in this case, the exploration, probably due to the hardest TTC constraint (i.e., greater imposed loads), is driven faster toward a better solution. Imposing 0.25* WCTTC, a dual PIC24 seems to solve the problem while, with 0.1* WCTTC, imposed loads are too high for SW implementations on available processors and so

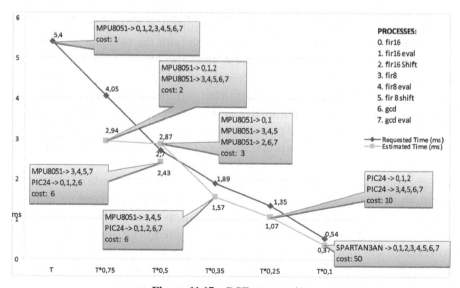

Figure 11.17 DSE step results.

a full HW architecture on FPGA is proposed. It is worth noting that, once a FPGA has been selected, one of the optimization goals is to exploit it to the best. So, given the very limited amount of HW resources occupied by the considered example, it can be fully contained on the FPGA and so the tool does not need to consider mixed HW/SW architectures.

All the steps, prior to DSE one, have been executed in approximately 30 min. It is worth noting that this is a one-time effort, while the described DSE step has been executed in less than 15 min by exploiting a *Microsoft Windows7*-based notebook equipped with an *Intel i7* processor and 4 GB of RAM. However, a not negligible amount of time has been wasted due to the "not so user-friendly" interface currently available for the prototypal co-design environment.

11.6 Conclusion

This Chapter has coped with the problem of the (electronic) system-level HW/SW co-design of dedicated digital systems based on heterogeneous multi-processor architectures. In particular, the work has considered an automatic system-level design space exploration (DSE) approach. On the basis of such a DSE approach, the work has presented a prototypal "SystemC-based Electronic System-Level HW/SW Co-Design Environment for Dedicated Heterogeneous Multi-Processor Systems". The presented tools are still in a prototypal status and several improvements are needed, but the preliminary experimental results are encouraging and justify further research efforts in this direction. In fact, the current work is oriented on both the two phases of the described DSE approach. For the first phase, since the *Affinity* metric has been exploited and improved by several works in the recent years (e.g., [163–165]), the goal is to integrate such improvements in the current methodology and also to introduce novel ones, e.g., to enrich the set of supported PUs (e.g., *Graphical Processing Unit*). For the second phase, the main goal is to develop the PAM2 tool by exploiting the design efforts already done for PAM1 and also to enhance the set of parameters and features used to characterize different ILs (e.g., latency, contention to access the IL, etc.). With respect to the whole co-design methodology, a full validation with respect to real-world data coming from lower levels of abstraction will be of very critical importance. Then, it will be mandatory to consider more non-functional requirements (i.e., power/energy) and to enhance the tools' user interface. Finally, it will be interesting also to exploit existing works to introduce in the presented SystemC-based co-design environment also the possibility of managing homogeneous/heterogeneous multi-core processors with shared memory among the cores [186].

References

[1] Edwards, M. (1997). "Software Acceleration using coprocessors: is it worth the effort?," in *Proceedings of 5th International Workshop on Hardware/Software Co-design* (Braunschweig: IEEE), 135–139.

[2] Amdahl, G. M. (1967). "Validity of single-processor approach to achieve large-scale computing capability," in *Proceedings of 30th AFIPS Spring Joint Computer Conference* (Reston, VA: AFIPS Press), 483–485.

[3] Axelsson, J. (1997). *Analysis and Synthesis of Heterogeneous Real-Time Systems*. Ph.D. thesis, Department of Computer and Information Science, Linköping University, Sweden.

[4] Allara, A., Fornaciari, W., Salice, F., and Sciuto, D. (1998). "A model for system-level timed analysis and profiling," in *Proceedings of the Design Automation and test in Europe, IEEE-DATE* (Paris: IEEE), 204–210.

[5] Kenter, H., Passerone, C., Smits, W., and Watanabe, Y. (1999). "Designing digital video systems: modeling and scheduling," in *Proceedings of the Seventh International Workshop on Hardware/Software Codesign, CODES, 1999*, Rome.

[6] Le Marrec, P., Valderrama, C., Hessel, F., Jerraya, A., Attia, M., and Cayrol, O. (1998). "Hardware, software and mechanical cosimulation for automotive applications," in *Proceedings of the Ninth International Workshop on Rapid System Prototyping*, Leuven, 202–206.

[7] Coste, P., Hessel, F., Marrec, P., Sugar, Z., Romdhani, M., Suescun, R., et al. (1999). "Multilanguage design of hererogeneous systems," in *Proceedings of the Seventh International Workshop on Hardware/Software Codesign, CODES '99* (Rome: IEEE), 54–58.

[8] Bolsen, I., De Man, H., Lin, B., Van Rompaey, K., Vercaiteren, S., and Verkest, D. (1997). Hardware/software co-design of digital telecommunication systems. *Proc. IEEE* 85, 391–418.

[9] Allara, A., Bombana, M., Fornaciari, W., and Salice, F. (2000). A case study in design space exploration: the TOSCA environment applied to a telecom link controller. *IEEE Des. Test Comput.* 17, 60–72.

[10] Brandolese, C. (2000). *A Co-Design Approach to Software Power Estimation for Embedded Systems*. Ph.D. thesis, Polytechnic University of Milan, Milano.

[11] Allara, A., Filipponi, S., Fornaciari, W., Salice, F., and Sciuto, D. (1997). "A flexible model for evaluating the behaviour of hardware/software systems," in *Proceedings of the Fifth International Workshop on Hardware/Software Codesign, CODES/CASHE '97* (Braunschweig: IEEE), 109–114.

[12] WoTUG (2015). *The Place for Communicating Processes*. Available at: http://www.wotug.org; *OCCAM Reference Manuals*. Available at: http://www.wotug.org/occam/documentation/

[13] Motorola (2015). *M68000 FAMILY Programmer's Reference Manual*. Available at: http://www.nxp.com/files/archives/doc/ref_manual/M680 00PRM.pdf

[14] IEEE Standard 1666-2011 (2011). *IEEE Standard 1666-2011: SystemC language*.

[15] Object Mangement Group (OMG) (2015). *Unified Modeling Language (UML) Resource Page*. Available at: http://www.uml.org/

[16] Schulz, S. E. (2000). "An introduction to SLDL and rosetta," in *Proceedings of the Asia and South Pacific Design Automation Conference, ASP-DAC* (Yokohama: IEEE), 571–572.

[17] Kamath, R., Alexander, P., and Barton, D. (1999). "SLDL: a systems level design language," in *Proceedings of the 12th Annual IEEE International ASIC/SOC Conference* (Washington, DC: IEEE), 71–75.

[18] Alexander, P., and Barton, D. (1999). "Rosetta: the SLDL constraints language," in *Proceedings of the Fall VIUF Workshop* (Rome: IEEE), 108–109.

[19] IEEE Standard 1076-1993 (1993). *IEEE Standard 1076-1993: VHDL Language Reference Manual*.

[20] ISO/IEC 8652:2012(E) (2012). *ISO/IEC Standard Ada Reference Manual – 8652:2012(E)*.

[21] Bal, H. E., Tanenbaum, A. S., and Kaashoek, M. F. (1990). ORCA: a language for distributed programming. *ACM SIGPLAN Notices* 25, 17–24.

[22] ISO/IEC 9899:2011 (2011). *Information Technology – Programming Languages – C*.

[23] Arnout, G. (1999). "C for system-level design," in *Proceedings of Design, Automation and Test Conference and Exhibition in Europe* (Munich: IEEE), 384–386.

[24] ISO/IEC 14882:2011(E) (2011). *Programming Languages — C++*, 3rd Edn.

[25] De Micheli, G. (1999). "Hardware synthesis from C/C++ models," in *Proceedings of Design, Automation and Test in Europe Conference and Exhibition* (Munich: IEEE), 382–383.

[26] Weiler, C., Kebschull, U., Rosensteil, W. (1995). "C++ base classes for specification, simulation and partitioning of a hardware/software system," in *Proceedings of the Asia and South Pacific Design Automation Conference* (Chiba: IEEE), 777–784.

[27] Verkest, D., Kunkel, J., and Schrirrmeister, F. (2000). "System-level design using C++," in *Proceedings of Design, Automation and Test Conference and Exhibition in Europe* (New York, NY: ACM), 74–81.

[28] Kuhn, T., and Rosenstiel, W. (2000). "Java based object oriented hardware specification and synthesis," in *Proceedings of the Asia and South Pacific Design Automation Conference, ASP-DAC* (Yokohama: IEEE), 579–581.

[29] Helaihel, R., and Olukotun, K. (1997). "Java as a specification language for hardware-software systems," in *Proceedings of the IEEE/ACM International Conference on Computer-Aided Design* (San Jose, CA: IEEE), 690–697.

[30] Lavagno, L., Pino, B., Reyneri, L.M., and Serra, A. (2000). "A simulink-based approach to system-level design and architecture selection," in *Proceedings of the 26th Euromicro Conference*, Vol. 1 (Maastricht: IEEE), 76–83.

[31] Berry, G., and Gonthier, G. (1992). The esterel synchronous programming language: design, semantics, implementation. *Sci. Comput. Program.* 19, 87–152.

[32] Mallet, F., and Boeri, F. (1999). "Esterel and Java in an object-oriented modelling and simulation framework for heterogeneous software and hardware systems," in *Proceedings of the 25th EUROMICRO Conference* (Milan: IEEE), 214–221.

[33] Boussinot, F., and De Simone, V. (1991). The Esterel language. *Proc. IEEE* 79, 1293–1304.

[34] Drusinsky, D., and Harel, D. (1989). "Using statecharts for hardware description and synthesis," in *Proceedings of the IEEE Transactions of CAD of Integrated Circuits and Systems*, Vol. 8 (Rome: IEEE), 798–807.

[35] Philipps, J., and Scholz, P. (1997). "Synthesis of digital circuits from hierarchical state machines," in *Proceeding of the Fifth GI/ITG/GMM Workshop*, Linz.

[36] Maciel, P., and Barros, E. (1996). "Capturing time constraints by using petri nets in the context of hardware/software codesign," in *Proceedings of the 7th IEEE International Workshop on Rapid System Prototyping* (Thessaloniki: IEEE), 36–41.

[37] Hoare, C. A. R. (2015). *Communicating Sequential Processes (2004 update)*. Available at: http://www.usingcsp.com/cspbook.pdf

[38] Harel, D., and Politi, M. (1998) *Modeling Reactive Systems with Statecharts: The Statemate Approach*. New York, NY; McGraw-Hill.

[39] Harel, D., Pnueli, A., Schmidta, J., and Sherman, R. (1987). On The Formal Semantics of Statecharts, in *"Proceedings of the Second IEEE Symposium on Logic in Computer Science,"* Ithaca, Greece.

[40] Gee, D. M., Worrall, J., and Henderson, W. D. (1994). *Using Statecharts in the Design of OCCAM 2 Programs*. Transputer Communications.

[41] Vahid, F., Narayan, S., Gajski, D. (1995). *A VHDL Front End for Embedded Systems*. IEEE Transactions on Computer Aided Design of Integrated Circuits and Systems, Vol. 14, No. 6, June 1995.

[42] Narayan, S., Vahid, F., and Gajski, D. (1991). System Specification and Synthesis with the SpecCharts Language, in *"Proceedings of the IEEE International Conference on Computer Aided Design"*.

[43] Narayan, S., Vahid, F., and Gajski, D. (1992). *System Specification with the SpecCharts Language*. IEEE Design & Test of Computers.

[44] GNU (2015). *Bison*. Available at: www.gnu.org/software/bison/bison.html

[45] Sima, D., Fountain, T., and Karsuk, P. (1997). *Advanced Computer Architectures: A Design Space Approach (International Computer Science Series)*, 1st Edn. Addison Wesley.

[46] Turner, K. J. (1993). *Using Formal Description Techniques*. Wiley.

[47] Armstrong, J., Virding, R., and Williams, M. (1993). *Concurrent Programming in Erlang*. Upper Saddle River, NJ: Prentice Hall.

[48] Zhu, J., Domer, R., and Gajski, D. (1997). "Syntax and Semantics of the SpecC Language," in *Proceedings of the SASIMI Workshop*, 75–82.

[49] Zhu, J., Domer, R., and Gajski, D. (1998). "IP-Centric Methodology and Design with the SpecC Language," in *Contribution to NATO-ASI Workshop on System-level Synthesis* (Barga, Italy).

[50] Zhu, J., Domer, R., and Gajski, D. (2000). *SpecC Specification Language and Methodology*. Norwell, MA: Kluwer Academic Publishers.

[51] Domer, R., and Gajski, D. (2000). "Reuse and Protection of Intellectual Property in the SpecC System," in *Proceedings of the Asia and South Pacific Design Automation Conference* (Tokyo: ASP-DAC), 49–54.

[52] Ku, D., and De Micheli, G. (1990). *HardwareC — A Language for Hardware Design (version 2.0 — 1990). SL Technical Report CSL-TR-90-419, Stanford University.* Available at: http://si2.epfl.ch/~demich el/publications/archive/1990/CSL-TR-90-419. pdf

[53] Hoffmann, D. W., Ruf, J., Kropf, T., and Rosenstiel, W. (2000). "Simulation Meets Verification-Checking Temporal Properties in SystemC," in *Proceedings of the 26th Euromicro Conference*, Vol. 1, ed. F. Vajda (Maastricht: Euromicro), 435–438.

[54] Narayan, S., and Gajski, D. D. (1993). "Features supporting system-level specification in HDLs," in *Proceedings European Conference on Design Automatio*n, (Hamburg), 540–545.

[55] Vahid, F., Gong, J., Gajski, D. D., and Nayaran, S. (1994). *Specification and Design of Embedded Systems.* Upper Saddle River, NJ: Prentice Hall.

[56] Rawson, J. (1994). "Hardware/Software Co-simulation," in *Proceeding 31st annual Design Automation Conference* (New York, NY: ACM), 439–440.

[57] Li, Y., and Malik, S. (1995). "Performance Analysis of Embedded Software Using Implicit Path Enumeration," in *Proceedings of the 32nd annual ACM/IEEE Design Automation Conference*, (New York, NY: ACM), 456–461.

[58] Suzuki, K., and Sangiovanni-Vincentelli A. *Efficient Software Performance Estimation Methods for Hardware/Software Co-design.* Proc. of DAC '96, Las Vegas, US.

[59] Lajolo, M., Lazarescu, M., Sangiovanni-Vincentelli, A. *A Compilation-based Software Estimation Scheme for Hardware/Software Co-Simulation.* Codes'99, Rome, Italy, pp. 85–89.

[60] Tabbara, B., Sgroi, M., Filippi, E., and Lavagno, L. (1999). *Fast Hardware–Software Co-simulation Using VHDL Models.* IEEE DATE.

[61] Carraras, C. et al. (1996). "A co-design methodology based on formal specification and high-level estimation," in *Proceeding of the IEEE Codes/CASHE'96*, (Pittsburgh, PA: IEEE).

[62] Fornaciari, W., Salice, F., and Sciuto, D. (1997). A Two-Level Cosimulation Environment. *IEEE Comput.* 30, 109–111.

[63] Hwang, K. (1979). *Computer Arithmetic: Principles, Architecture, and Design.* New York, NY: J. Wiley and Sons.

[64] Gajski, D., Dutt, N., Wu, A., and Lin, S. (1992). *High-level Synthesis.* Norwell, MA: Kluwer Academic Publisher.

[65] Meinel, C., and Theobald, T. (1998). *Ordered Binary Decision Diagrams and their Significance in Computer-aided Design of VLSI Circuits – a Survey. Electronic Colloquium on Computational Complexity (ECCC)*, Report n. 39.

[66] Coumeri, S. L. and Thomas, D. (1995). "A simulation environment for hardware-software codesign," in *Proceedings of IEEE International Conference on Computer Design*, (Austin, TX: IEEE), 58–63.

[67] Marwedel, P. (1990). "Matching system and component behavior in MIMOLA synthesis tools" in *Proceedings of the European Design Automation Conference* (Los Alamitos, CA: IEEE Computer Society Press), 146–156.

[68] Marwedel, P. (1985). "The mimola design system: a design system which spans several levels," in *Methodologies of Computer System Design*, ed. B. D. Shriver (North Holand), 223–237.

[69] Fauth, A., Van Praet, J., and Freericks, M. (1995). "Describing instruction set processors using nML," in *Proceedings of the European Design and Test Conference*, (Paris: IEEE), 503–507.

[70] Rajesh, V., and Moona, R. (1999). "Processor modeling for hardware software codesign," in *Proceedings of the 12th International Conference on VLSI Design*, (Goa: IEEE), 132–137.

[71] Ptolemaeus, C. (Editor) (2014). *System Design, Modeling, and Simulation Using Ptolemy II*. Available at: http://Ptolemy.org

[72] Liem, C., Nacabal, F., Valderrama, C., Paulin, P., and Jerraya, A. (1997). *System-on-a-chip Co-simulation and Compilation. IEEE Design Test Comput.* 14, 16–25.

[73] Balarin, F., Giusto, P., Jurecska, A., Passerone, C., Sentovich, E., Tabbara, B., et al. (1997). *Hardware-Software Co-Design of Embedded Systems — The POLIS Approach. The Springer International Series in Engineering and Computer Science*. Available at: http://embedded.eecs.berkeley.edu/research/hsc/

[74] Van Rompaey, K., Bolsens, I., De Man, H., and Verkest, D. (1996). "CoWare — a design environment for heterogenous hardware/software systems," in *Proceedings of the Conference on European Design Automation (EURO-DAC '96/EURO-VHDL '96)* (Los Alamitos, CA: IEEE Computer Society Press).

[75] Mentor Graphics Seamless. (2015). http://www.mentor.com/products/fv/seamless/

[76] Intel. (2015). *Embedded Ultra Low Power Intel 486 GX Processor Datasheet*, Available at: http://media.digikey.com/pdf/Data%20Sheets/Intel%20PDFs/Intel486_GX_Dec97.pdf

[77] Jonsson, J., and Vasell, J. (2015). *A Preliminary Survey on Predictable Fine-Grain Parallel Computer Architectures*, Technical Report No.180, Goteborg: Chalmers University of Technology. Available at: http://libris.kb.se/bib/1808520

[78] Stankovic, J. A., and Ramamritham, K. (1988). *Tutorial: Hard Real-Time Systems*. (Los Alamitos, CA: Computer Society Press), 1–11.

[79] Taylor, C. E., Schroeder, R. N. (1995). "Today's RISC microprocessors and real-time concerns," in *Proceeding of the ICS/95 the Industrial Computing Society Conference*.

[80] Johansson, R. (1992). *Processor Performance in Real-Time Systems*, Licentiate thesis No. 136L, Department of Computer Engineering, Chalmers University of Technology, Goteborg.

[81] Hand, T. (1989). *Real-Time Systems Need Predictability*. Computer Design RISC Supplement.

[82] Simpson, D. (1989). Real-time RISCs. *Syst. Integr.* 35–38.

[83] Claasen, T. A. C. M. (1999). "High speed: not the only way to exploit the intrinsic computational power of silicon," in *Proceeding of the IEEE International Solid-State Circuits Conference*, 22–25.

[84] Sander, G., Tamassia, R., and Tollis, I. G. (1994). "Graph layout through the VCG tool," in *Proceedings of DIMACS International Workshop GD'94*, eds R. Tamassia, and I. G. Tollis, (Princeton, NJ: Springer-Verlag Berlin Heidelberg).

[85] Choi, J., Burke, M., and Carini, P. (1993). "Efficient flow-sensitive interprocedural computation of pointer-induced aliases and side effects," in *Proceedings of the 20th Annual ACM Symposium on Principles of Programming Languages*, 233–245.

[86] Brown, S. D., Francis, R. J., Rose, J., and Vranesic, Z. G. (1992). Field-programmable gate arrays. series the springer international series in Engineering and Computer Science. 180.

[87] Kilts, S. (2007). *Advanced Fpga Design: Architecture, Implementation, and Optimization*. Hoboken, NJ: Wiley-Interscience.

[88] Arifur, R., and Anderson, J. H. (eds). (2015). "FPGA based design and applications," in *Integrated Circuits and Systems*. Berlin: Springer.

[89] De Hon, A. (1996). *Reconfigurable Architectures for General-Purpose computing*, Technical Report 1586, MIT-AI Laboratory. Available at: http://www.ece.iastate.edu/~zambreno/classes/cpre583/documents/Deh96A.pdf

[90] Cooper, T. (2015). *Taming the SHARC*. Technical report, Ixthos Inc., 2000. Available at: http://www.eetasia.com/ARTICLES/2001SEP/2001SEP26_DSP_AN.PDF?SOURCES=DOWNLOA

[91] Analog Devices Inc. (2004). *ADSP-2106x SHARC Processor User's Manual Revision 2.1.*

[92] Knuth, D. (1971). An empirical study of FORTRAN programs. *Softw. Pract. Exp.* 1, 105–1333.

[93] Kung, S.Y. (1988). *VLSI Array Processors.* Upper Saddle River: Prentice Hall.

[94] Guerra, L., Potkonjak, M., Rabaey, M. (1994). "System-level design guidance using algorithm properties," in *Proceeding of the J. VLSI Signal Processing Workshop on VII*, 73–82.

[95] Kavalade, A., and Lee, A. (1994). "A global criticality/local phase driven algorithm for the constrained hardware/software partitioning problem," in *Proceedings of Codes/CASHE '94, Third Intl. Workshop on Hardware/Software Codesign*, Grenoble, 22–24.

[96] Carro, L., Kreutz, M., Wagner, F. R., and Oyamada, M. (2000). "System synthesis for multi-processor embedded applications," in *Proceedings of the Design, Automation and Test in Europe Conference and Exhibition 2000*, 697–702.

[97] Salice, F., Fornaciari, W., Sciuto, D. (2002). *Conceptual-Level Hw/Sw Co-Design of Embedded Systems: Partitioning of UML Specifications for Hw/Sw Architectures.*

[98] Devanbu, P. (1992). GENOA: a customizable, language- and front-end independent code analyzer. *Proc. ICSE* 92, 307–317.

[99] Jifeng, H., Page, I., Bowen, J. (1993). "Towards a provably correct hardware implementation of occam," in *Proceedings of the IFIP WG 10.5 Advanced Research Working Conference on Correct Hardware Design and Verification Methods CHARME '93*, Berlin: Springer-Verlag.

[100] Ronngren, S., and Shirazi, B. A. (1995). Static multi-processor scheduling of periodic real-time tasks with precedence constraints communication costs, in *Proceeding of the of 28th Annual Hawaii IEEE International Conference on System Sciences*, 143–152.

[101] Bjorn-Jorgensen, P., and Madsen, J. (1997). Critical path driven cosynthesis for heterogeneous target architectures, in *Proceeding of the 5th IEEE/ACM International Workshop on Hardware/Software Co-Design CODES/CASHE'97*, Braunscheig.

[102] Sih, G. C., and Lee, E. A. (1993). "A compile-time scheduling heuristic for interconnection-constrained heterogeneous processor architectures," in *Proceeding of the IEEE Transaction On Parallel and Distributed Computing*, 4, 175–187.

[103] Rhodes, D. L. (1999). *Real-Analysis, ALAP Guided Synthesis.* Ph.D. thesis, Princeton University, Princeton, NJ.

[104] Rhodes, D. L., Wolf, W. (1999). *Computer-Aided Design,* Digest of Technical Papers. IEEE/ACM International Conference, 339–342.

[105] Gupta, R., Spezialetti, M. (1996). A compact task graph representation for real-time scheduling. *Real Time Syst.* 11, 71–102.

[106] Synopsys. (2015). *Design Compiler,* Available at: http://www.synopsys. com/Tools/Implementation/RTLSynthesis/Pages/default.aspx

[107] Strik, M. T. J., Timmer, A. H., Van Meerbergen, J. L., Van Rootselaar, G. (2000). Heterogeneous multi-processor for the management of real-time video and graphics streams. *Solid State Circ. IEEE J.* 35, 1722–1731.

[108] Hilgenstock, J., Herrmann, K., Moch, S., Pirsch, P. (2000). A single-chip video signal processing system with embedded DRAM. *Signal Proc. Syst.* 23–32.

[109] Ixthos. (2015). *Common Heterogeneous Architecture for Multi-Processing (CHAMP),* Available at: http://www.eetkorea.com/ARTICLES/20 01SEP/2001SEP25_MPR_DSP_AN.PDF?SOURCES=DOWNLOAD

[110] www.mc.com. (2015).

[111] www.alacron.com. (2015).

[112] www.skycomputers.com, (2015)

[113] Ernst, R., Henkel, J., Benner, T. H., Ye, W., Holtmann, U., Herrmann, D., and Trawny, M. (1996). The COSYMA environment for hardware/ software cosynthesis of small embedded systems, Microprocessors and Microsystems, 20, 159–166.

[114] Knieser, M. J., and Papachriston, C. A. (1996). "COMET: a hardware-software codesign methodology," in *Design Automation Conference, 1996, with EURO-VHDL '96 and Exhibition, Proceedings EURO-DAC '96, European* (178–183).

[115] Niemann, R. (1998). *Hardware/Software Co-Design for Data Flow Dominated Embedded Systems.* Kluwer Academic Publishers.

[116] Ismail, T. B., Abid, M., and Jerraya, A. (1994). "*COSMOS:* a codesign approach for communicating systems," in *Proceedings of the Third International Workshop on Hardware/Software Codesign,* (IEEE Computer Society Press), 17–24.

[117] Chou, P. H., Ortega, R. B., and Borriello, G.(1995). "The Chinook hardware/software co-synthesis system," in *Proceedings of the Eighth International Symposium on System Synthesis* (Cannes, IEEE), 22–27.

[118] Tsasakou, S., Voros, N. S., Koziotis, M., Verkest, D., Prayati, A., and Birbas, A. (1999). "Hardware-software co-design of embedded systems

using CoWare's N2C methodology for application development," in *Proceedings of the ICECS '99. The 6th IEEE International Conference on Electronics, Circuits and Systems* (Pafos, IEEE). 1, 59–62.

[119] Dave, B. P., Lakshminarayana, G., and Jha, N. K. (1999). "COSYN: hardware-software co-synthesis of heterogeneous distributed embedded systems," in *Proceedings of the Very Large Scale Integration (VLSI) Systems, IEEE Transactions on* (IEEE), 7, 92–104.

[120] Dave, B. P., and Jha, N. K. (1998). "COHRA: hardware-software cosynthesis of hierarchical heterogeneous distributed embedded systems," in *Proceedings of the Computer-Aided Design of Integrated Circuits and Systems, IEEE Transactions on* (IEEE) 17, 900–919.

[121] Hsiung, P. A. (2000). "CMAPS: a cosynthesis methodology for application-oriented general-purpose parallel systems," in *Proceedings of the ACM Transactions on Design Automation of Electronic Systems* (New York, NY, ACM), 5, 58–81.

[122] The Mathworks (2015). *The Mathworks Products*. Available at: http://www.mathworks.com/products

[123] Janka, R. S., and Wills, L. (2000). "Combining virtual benchmarking with rapid system prototyping for real-time embedded multi-processor signal processing system codesign," in *Proceedings of the Rapid System Prototyping, 2000 11th International Workshop on* (IEEE), 20–25.

[124] Microsoft (2015). *Microsoft Excel*. Available at: www.microsoft.com

[125] Lyonnard, D., Sungjoo, Y., Baghdadi, A., and Jerraya, A. A. (2001). "Automatic generation of application-specific architectures for heterogeneous multi-processor system-on-chip," in *Proceedings of the Design Automation Conference, 2001* (IEEE), 518–523.

[126] IEEE (2005). *1364–2005: Verilog Hardware Description Language*.

[127] Ernst, R., Henkel, J., and Benner, T. (1993). *Hardware-Software Cosynthesis for Microcontrollers*. IEEE Design & Test of Computers (64–75).

[128] Gupta, R. K., and De Micheli, G. (1992). "System-level synthesis using re-programmable components," in *Proceedings of the IEEE European Conference on Design Automation*, Washington, 2–7.

[129] De Man, I., Bolsen, B., Lin, K., Van-Rompaey, S., Vercauteren, D., and Verkest, D. (1995). "Co-design for DSP systems," in *Proceedings of the NATO ASI Hardware/Software Co-design*, Tremezzo.

[130] Marchioro, G. F., Daveau, J. M., and Jerraya, A. A. (1997). "Transformational partitioning for co-design of multi-processor systems," in *Proceedings of the IEEE ICCAD '97* (San Jose, California: IEEE), 9–13.

[131] Srivastava, M., and Brodersen, R. (1995). "SIERA: a unified framework for rapid-prototyping of system-level hardware and software," in *Proceedings of the IEEE Transactions on Computer-Aided Design of Integrated Circuits and Systems (IEEE)*, 676–693.

[132] Kavalade, A., and Lee, E. A. (1995). "The extended partitioning problem: hardware/software mapping, scheduling, and implementation-bin selection". *in Proceedings of the Sixth Workshop on Rapid Systems Prototyping* (IEEE), 12–18.

[133] Paulin, P., Liem, C., May, T., and Sutarwala, S. (1994). "DSP design tool requirements for embedded systems: a telecommunication industrial perspective," in *Proceedings of the Journal of VLSI Signal Processing (special issue on synthesis for real-time DSP)* (Kluwer Academic Publisher) 1994.

[134] Agrawal, S., and Gupta, R. K. (1997). "Data-flow assisted behavioral partitioning for embedded systems" in *Proceedings of the 34th Conference on Design Automation (DAC97)* (NY: ACM Press) 709–712.

[135] Knudsen, P. V., and Madsen, J. (1999). *"Graph based communication analysis for hardware/software codesign,"* in *Proceedings of the Hardware/Software Codesign, (CODES '99) Seventh International Workshop on* (Rome: IEEE), 131–135.

[136] Chang, J. M., and Pedram, M. (2000). *"Codex-dp: co-design of communicating systems using dynamic programming,"* in *Proceedings of the Computer-Aided Design of Integrated Circuits and Systems, IEEE Transactions on* (IEEE), 19, 732–744.

[137] Hidalgo, J. I., and Lanchares, J. (1997) *"Functional partitioning for hardware-software codesign using genetic algorithms,"* in *Proceedings of the 23rd EUROMICRO Conference EUROMICRO 97. New Frontiers of Information Technology* (Budapest: IEEE) 631–638.

[138] Adams, J. K., and Thomas, D. E. (1995) *"Multiple-process behavioral synthesis for mixed hardware-software systems,"* in *Proceedings of the Eighth International Symposium on* (Cannes: IEEE), 10–15.

[139] Mitchell, M. (1996). *An Introduction to Genetic Algorithms*, (M.I.T. Press) 1996.

[140] Matthew Wall using (2015). *GALIB*. Available at: http://lancet.mit.edu/ga/

[141] Liu, C. L., and Layland, J. W. (1973). scheduling algorithms for multiprogramming in a hard real-time environment, *Journal of the ACM*, 20, 37–53.

[142] Axelsson, J. (1999). "Towards system-level analysis and synthesis of distributed real-time systems," in *Proceedings of the 5th International Conference on Information Systems Analysis and Synthesis* Orlando, FL, 5, 40–46.

[143] Xilinx (2015). *Virtex-II Pro.* Available at: http://www.xilinx.com

[144] NXP Semiconductor. (2015). *Nexperia Platform.* Available at: www.nxp.com

[145] Texas Instruments. (2015). *OMAP Platform.* Available at: http://www.ti.com/lsds/ti/processors/dsp/c6000_dsp-arm/omap-11x/overview.page

[146] Renesas. (2015). *SH Mobile Series.* Available at: http://www.renesas.com/products/soc/assp/mobile/index.jsp

[147] Auguin, M., Capella, L., Cuesta, F., and Gresset, E. (20001). "CODEF: a system level design space exploration tool," in *Proceedings of the IEEE Int. Conference on Acoustics, Speech, and Signal Processing* (Salt Lake City, UT: IEEE), 2, 1145–1148.

[148] Baghdadi, A., Zergainoh, N. E. Cesario, W. O., and Jerraya, A. A. (2002). Combining a performance estimation methodology with a hardware/software codesign flow supporting multiprocessor systems. *IEEE Trans. on SW Engineering*, 28, 822–831.

[149] Balarin, F., Watanabe, Y., Hsieh, H., Lavagno, L., Passerone C., and Sangiovanni-Vincentelli, A. (2003). Metropolis: An integrated electronic system design environment. *IEEE Comput.* 36, 45–52.

[150] Pomante, L., Sciuto, D., Salice, F., Fornaciari, W., Brandolese, C. (2006). Affinity-driven system design exploration for heterogeneous multiprocessor SoC. *IEEE Trans. Comput.* 55, 5.

[151] Streicher, T., Glab, M., Haubelta C., and Teich J. (2007). Design space exploration of reliable networked embedded systems. *J. Syst. Architect.* 53, 751–763.

[152] Ascia, G., Catania, V., Di Nuovo, A. G., Palesi, M., and Patti, D. (2007). Efficient design space exploration for application specific systems-on-a-chip. *J. Syst. Archit.* 53, 733–750.

[153] Holzer, M., Knerr, B., and Rupp, M. (2007). "Design space exploration with evolutionary multiobjective optimization," in *Proceedings of the International Symposium on Industrial Embedded Systems*, Lisbon, 125–133.

[154] Palermo, G., Silvano, C., and Zaccaria, V. (2008). "An efficient design space exploration methodology for on-chip multiprocessors subject to application specific constraints," in *Proceedings of the Symposium on Application Specific Processors, SASP 2008*, Anaheim, CA, 75–82.

[155] Haubelt, C., Schlichter, T., Keinert, J., and Meredith, M. (2008). "SystemCo-designer: automatic design space exploration and rapid prototyping from behavioral models," in *Proceedings of the 45th ACM/IEEE on Design Automation Conference,* Anaheim, CA, 580–585.

[156] Anderson, I. D. L., and Khalid, M. A. S. (2009). SC Build: A computer-aided design tool for design space exploration of embedded central processing unit cores for field-programmable gates arrays. *Inst. Eng. Technol. Comput. Digital Tech.* 3, 24–32.

[157] Bailey, B., Martin, G., Piziali, A. (2007). *ESL Design and Verification: A Prescription for Electronic System Level Methodology.* Burlington: Morgan Kaufman Publishers.

[158] Vahid, F., and Givargis, T. (2002). *Embedded System Design: A Unified HW/SW Introduction.* Hoboken: John Wiley & Sons.

[159] Hoare, C. A. R. (1978). Communicating sequential processes. *Commun. ACM* 21, 666–676.

[160] Teich, J., Blickle, T., and Thiele, L. (1997). "An evolutionary approach to system-level synthesis," in *Proceedings of the Fifth International Workshop on Hardware/Software Codesign, (CODES/CASHE '97),* Braunschweig, 167–171.

[161] Sciuto, D., Salice, F., Fornaciari, W., and Pomante, L. (2001). Hw/Sw Co simulation for Fast Design Space Exploration of Multiprocessor Embedded Systems. *Can. J. Elect. Comput. Eng.* 26, 135–140.

[162] Taddei, P., and Tornatore, A. (2003). *Sched_PA: a scheduler in SystemC,* Master Thesis, Master of Science in Electrical Engineering and Computer Science, University of Illinois, Chicago.

[163] Issam, M., Guy, G., Mohamed, A., and Luc, P. J. (2004). "Metrics for multiprocessor system on chip," in *Proceedings of the 16th International Conference on Microelectronics.*

[164] Abildgren, R., Saramentovas, A., Ruzgys, P., Koch, P., and Le Moullec y. (2007). "Algorithm-architecture affinity-parallelism changes the picture," in *Proceedings on the Design and Architectures for Signal and Image Processing,* Grenoble, France.

[165] Dey, S., Kedia, M., and Basu, A. (2008). "An Approach to Software Performance Evaluation on Customized Embedded Processors," in *Proceedings of the 21st International Conference on VLSI Design.*

[166] Pomante, L. (2011). "System-level design space exploration for dedicated heterogeneous multi-processor systems," in *Proceedings of the IEEE International Conference on Application-specific Systems, Architectures and Processors,* Washington, DC.

[167] Allara, A., Brandolese, C., Fornaciari, W., Salice, F., and Sciuto, D. (1998). "System-level performance estimation strategy for sw and hw," in *Proceedings of the International Conference on Computer Design: VLSI in Computers and Processors ICCD '98*, 48–53.

[168] Brandolese, C., Fornaciari, W., and Salice, F.(2004). "An area estimation methodology for FPGA based designs at systemc-level," in *Proceedings of the 41st Design Automation Conference*, San Diego, CA, 129–132.

[169] Brandolese, C. (2008). "Source-level estimation of energy consumption and execution time of embedded software," in *Proceedings of the 11th EUROMICRO Conference on Digital System Design Architectures, Methods and Tools, DSD '08, Parma*, 115–123.

[170] Iyoda, J., Sampaio, A., and Silva, L. (1999). "ParTS: a partitioning transformation system," in *Proceedings of the Wold Congress on Formal Methods in the Development of Computing Systems-Volume II (FM '99)*, Springer-Verlag, London, 1400–1419.

[171] DK Design Suite (2015). *DK Design Suite*. Available at: https://www.me ntor.com/products/fpga/handel-c/dk-design-suite

[172] ImpulseC (2015). *ImpulseC*. Available at: http://www.impulseaccelerat ed.com

[173] Claus, B., and Wolfgang, N. (2007). "CSP with Synthesizable SystemC(tm) and OSSS," in *Proceedings of FDL 2007*.

[174] Patel, H. D., Mathaikutty, D., and Shukla, S. K. (2004). *Implementing Multi-Moc Extensions for SystemC: Adding CSP and FSM Kernels for Heterogeneous Modelling*. Technical Report, FERMAT Research Laboratory, Virginia Tech.

[175] Herrera, F., Sanchez, P., and Villar, E. (2004). "Modeling of CSP, KPN and SR systems with systemC," in *Languages for System Specification*, ed. Grimm C(Norwell, MA: Kluwer Academic Publishers), 133–148.

[176] David, C., Ku, and De Micheli, G. (1992). *High Level Synthesis of Asics Under Timing and Synchronization Constraints*. Norwell, MA: Kluwer Academic Publishers.

[177] Gupta, R. K. (1995). *Co-Synthesis of Hardware and Software for Digital Embedded Systems*. Norwell, MA: Kluwer Academic Publishers.

[178] Xilinx Zynq7000 (2015). *Xilinx Zynq7000*. Available at: http://www.xili nx.com

[179] OMAP Platform (2015). *OMAP Platform*. Available at: www.omap.com

[180] SH Mobile Series (2015). *SH Mobile Series*. Available at: http://www.renesas.com

[181] Belwal, M., and Sudarshan, T. S. B. (2014). "A survey on Design Space Exploration for Heterogeneous Multi-core," in *Proceedings of the International Conference on Embedded Systems 2014 (ICES 2014)*, Coimbatore.

[182] Jia, Z. J., Bautista, T., Núñez, A., Pimentel, A. D., and Thompson, M. (2013). A system-level infrastructure for multidimensional MP-SoC design space co-exploration. ACM Trans. *Embedd. Comput. Syst.* 13, 26.

[183] Pomante, L., Serri, P., Incerto, E., and Volpe, J. (2014). "HW/SW Co-design of heterogeneous multiprocessor dedicated systems: a systemC-based environment," in *Proceedings of the second World Congress on Multimedia and Computer Science International Conference on Telecommunications and Modelling, Analysis and Simulation of Computer Systems (ICTMASCS'2014)*, Hammamet, Tunisia.

[184] Page, I. (1996). "Closing the gap between hardware and software: hardware-software cosynthesis at Oxford," in *Proceedings of the IEE Colloquium on Hardware-Software Co synthesis for Reconfigurable Systems (Digest No: 1996/036)*, 2/1–211.

[185] http://www.eembc.org/

[186] Pomante, L. (2013). "HW/SW Co-Design of Dedicated Heterogeneous Parallel Systems: an Extended Design Space Exploration Approach," in *Proceedings of the IET Computers and Digital Techniques, Institution of Engineering and Technology*, 246–254.

[187] Intel CoFluent (). *Intel CoFluent*. Available at: www.intel.com

[188] Keutzer, K., Malik, S., Newton, A., Rabaey, J., and Sangiovanni-vincentelli, A. (2000). "System level design: orthogonalization of concerns and platform-based design," in *Proceedings of the IEEE Transactions on Computer-Aided Design of Integrated Circuits and Systems*, 1523–1543.

[189] Cai, L., and Gajski, D. (2003). "Transaction level modeling: an overview," in *Proceedings of the First IEEE/ACM/IFIP International Conference on Hardware/Software Codesign and System Synthesis*, 2003, pp.19–24.

Index

A

Abstract 7, 38, 155, 177
Abstraction 13, 37, 142, 183
Access 31, 35, 77, 222
Accuracy 16, 48, 87, 141
Accurate 5, 48, 71, 138
Active 146, 171
Activity 13, 141, 205, 219
Address 22, 68, 87, 106
Affinity 79, 83, 127, 162
Allocation 30, 133, 213, 214
Algorithm 120, 146, 181, 205
Analysis 4, 19, 63, 160
Analyze 6, 20, 85, 183
Annotation 23, 37, 38, 39
Application 23, 97, 156, 166
Application-specific 7, 15, 43, 123
Approach 27, 70, 126, 142
Approximately 90, 128, 168, 222
Architectural 13, 66, 69, 208
Architecture 4, 44, 185, 209
Architecture graph 186, 187, 206
Area 31, 39, 53, 202
Arithmetic 52, 99, 100, 110
Array 66, 73, 75, 108
Arrow 60, 183
ASIC 69, 78, 129, 202
Aspect 44, 87, 89, 150
Assembly 29, 35, 94, 117
Assignment 20, 98, 119, 144
Assumption 20, 94, 101, 150
Automatic 19, 171, 208, 216
Automatically 30, 183, 200, 202
Available 16, 70, 141, 186
Average 59, 71, 83, 151
Avoid 31, 112, 154, 209

B

Back annotation 37, 38, 39, 143
Bandwidth 65, 179, 186, 192
Basic 30, 53, 57, 186
Basic block 185, 186, 205
BB 185, 188, 193, 208
Behavior 16, 39, 201, 206
Behavioral 14, 39, 126, 200
Bit 29, 53, 69, 79
Block 53, 89, 167, 186
Body 55, 58, 67, 104
Boolean 52, 70, 101, 108
Branch 36, 94, 103, 112
Buffer 68, 148, 167, 182
Build 34, 55, 130, 203

C

Call 59, 97, 150, 156
Case study 159, 165, 216, 219
Channel 31, 115, 149, 207
Characteristic 23, 51, 93, 158
Characterization 65, 87, 124, 141
Characterize 63, 86, 141, 198
Check 39, 167, 199, 206
Child 55, 110, 112, 116
Children 111, 112, 113, 116
Chip 69, 175, 181, 199
Circular 68, 75
Class 29, 53, 58, 117
Classical 4, 49, 133, 202
Classification 73, 127, 175
Clock 49, 91, 111, 204
Cluster 132, 133, 164, 169
Clustering 5, 126, 133, 202
Co-design 1, 170, 177, 201
Co-simulation 139, 162, 167, 203

239

Coarse 18, 48, 126, 202
Code 29, 37, 89, 103
Coefficient 80, 146, 151
Collection 24, 30, 37, 117
Combination 21, 33, 99, 130
Combinatorial 33, 109, 111, 116
Commercial 7, 49, 122, 201
Common 29, 52, 175, 193
Communication 4, 104, 114, 129
Compact 47, 74, 144, 163
Compilation 31, 33, 34, 96
Compiler 35, 87, 105, 171
Complex 21, 70, 121, 199
Complexity 8, 70, 78, 209
Component 10, 142, 176, 199
Computation 70, 109, 177, 204
Computational 8, 57, 74, 198
Concept 12, 44, 91, 203
Conclusion 25, 83, 138, 198
Concurrency 52, 70, 190, 204
Concurrent 13, 77, 186, 193
Condition 52, 102, 103, 112
Connected 15, 43, 99, 208
Constant 70, 75, 76, 98
Constraint 137, 154, 188, 208
Container 29, 53, 75, 108
Context 71, 146, 183, 210
Contribution 80, 98, 145, 171
Control 33, 91, 103, 151
Correctness 9, 39, 143, 216
Cost 10, 130, 137, 186
Co-synthesis 20, 21, 23, 24
Couple 96, 102, 194
CPI 91, 92, 95, 118
Critical 10, 46, 125, 206
Crossover 133, 190,
    191, 209
CSP 184, 190, 207, 217
CU 194, 195, 205
Cycle 48, 87, 112, 141

D
DAG 68, 99, 109
Data 37, 73, 118, 190
Datum 91, 106
Dead 15, 36, 139, 162

Decision 74, 101, 102
Declaration 53
Decomposition 89, 125, 155, 202
Dedicated 31, 175, 199, 209
Definition 57, 72, 127, 209
Degree 39, 76, 90, 123
Delay 53, 92, 127, 217
Dependent 13, 78, 178, 216
Derived 32, 37, 120, 183
Description 17, 44, 89, 216
Design 1, 13, 187, 199
Design space exploration 15, 175,
    187, 199
Designer 12, 133, 189, 192
Detail 16, 47, 127, 206
Detect 4, 39, 114, 167
Detection 42, 116, 178, 204
Determine 34, 79, 93, 188
Developed 29, 72, 120, 199
Development 4, 30, 65, 210
Device 65, 71
Difference 37, 52, 71, 176
Different 8, 83, 141, 190
Digital 7, 65, 175, 199
Digital signal processing 8, 65, 82
Directives 30, 46, 140, 204
Directly 37, 98, 193, 207
Discrete 22, 25, 146, 199
Distributed 3, 21, 150, 205
DSE 188, 204, 209, 220
DSP 66, 75, 83, 190
Dual 67, 158, 221
Dynamic 28, 121, 122, 168

E
Early 17, 41, 124, 178
Edge 111, 113, 116, 192
Effective 10, 83, 141, 201
Effectiveness 5, 44, 155, 170
Effect 97, 107
Efficiency 5, 17, 71, 141
Efficient 12, 23, 70, 125
Effort 3, 23, 143, 203
Electronic 22, 176, 199, 207
Electronic system level 200, 206,
    216, 222

Element 4, 39, 117, 144
Embedded 1, 7, 21, 181
Embedded system 8, 13,
    16, 26
Enabling 3, 48, 81, 171
Engine 85, 135, 143, 215
Entity 33
Entry 24, 51, 153, 190
Environment 27, 39,
    141, 208
Equal 9, 101, 156, 220
Equivalent 16, 81,
    179, 202
Errors 11, 121, 122, 180
ESL 199, 201, 206, 208
Essential 9, 85, 106, 142
Estimate 5, 39, 143, 204
Estimation 4, 41, 87, 207
Evaluate 20, 79, 128, 178
Evaluation 5, 107, 151, 179
Evolution 133, 189, 194, 210
Example 70, 120, 155, 195
Executable 19, 54, 117, 200
Execution 52, 65, 107, 144
Executor 71, 79, 128, 130
Exist 34, 50, 58, 176
Experimental 64, 86,
    157, 198
Explicit 8, 49, 160, 189
Exploit 13, 70, 188, 189

F
Factor 9, 80, 120, 175
Feasible 24, 39, 151, 186
Feature 41, 64, 134, 175
FIFO 150, 206
File 34, 117, 118, 164
Flexible 29, 70, 142, 182
Flow 32, 37, 165, 167
Focus 6, 44, 108, 195
Formal 14, 33, 72, 180
Formalism 21, 32, 183, 202
FPGA 129, 137, 164, 202
Framework 14, 19, 46, 117
Function 81, 130, 137, 193
Functional 152, 162, 167, 207

Functional co-simulation 42, 152,
    162, 167
Functionality 4, 39, 66, 128

G
Gate 7, 120, 141, 179
Generic 76, 90, 95, 194
Genetic 20, 125, 193, 209
GPP 66, 78, 157, 189
Granularity 23, 71, 126, 165
Graph 132, 156, 166, 192
Graphical 34, 60, 175, 199
Groups 54, 71, 131, 132

H
Hierarchy 9, 50, 52, 89
High-level 39, 64, 124, 140
High-level synthesis 49, 85, 175
Homogeneous 39, 48, 135, 222
Hop 198

I
Idea 63, 83, 85, 107
Ideal 14, 70, 128, 163
Identification 5, 70, 140, 204
IF 29, 56, 102, 112
IL 95, 189, 192, 193
Imperative 54, 89, 153, 182
Implement 33, 70, 129, 205
Implementation 11, 74, 105, 207
Improvements 17, 50, 198, 222
Independent 14, 29, 103, 183
Individual 133, 191, 194, 197
Information 16, 30, 39, 153
Innovative 4, 41, 81, 172
Input 53, 105, 108, 149
Instance 59, 128, 193, 220
Instruction 71, 97, 117, 175
Integer 66, 69, 73, 79
Interconnection 45, 134, 157, 201
Intermediate representation 4
Internal 31, 54, 57, 60
Internal model 55, 57, 60, 183
Issue 11, 126, 176, 213
Item 133, 180, 189, 205
Iteration 104, 137, 164, 169

K

Kernel 5, 30, 117, 207

L

Language 28, 51, 94, 183
Length 66, 68, 69, 101
Level 3, 9, 13, 16
Library 19, 22, 33, 41
Limitation 20, 43, 48, 109
Load 17, 30, 42, 81
Load estimation 42, 152, 162, 168
Logic 33, 66, 90, 101
Loop 3, 13, 36, 52
LSI 120, 122

M

MAC 68, 75, 79, 80
Main 4, 7, 13, 30
Management 10, 65, 78, 151
Manager 66, 149, 151, 216
Manipulation 70, 78, 155
Manually 16, 47, 124, 181
Mapping 18, 36, 94, 126
Matching 16, 44, 72, 80
Matrix 25, 75, 193, 195
Max 53, 130, 185
Maximum 44, 94, 100, 129
Media 4, 39, 70, 151
Memory 7, 24, 67, 77
Methodology 3, 24, 43, 72
Method 58, 118, 125, 183
Metric 71, 80, 185, 202
Microprocessor 8, 48, 87, 141
Min 53
Minimize 8, 19, 96, 124
Minimizing 18, 131, 189, 209
Minimum 95, 100, 119, 164
Mixed 13, 28, 51, 141
Model 4, 22, 32, 57
MPU 2, 19, 221
Multi processor 12, 134, 146, 171
Multimedia 13, 141, 156

N

Network 7, 46, 112, 151
Node 12, 55, 99, 183

Normalization 79, 80, 81
Notation 47, 186

O

Object 22, 54, 117, 143
OCCAM 29, 34, 51, 53
Operand 99, 100, 111
Operating 30, 66, 120, 146
Operator 92, 99, 109, 119
Optimal 12, 25, 44, 57
Optimization 5, 36, 90, 125
OSTE 120, 152
Output 29, 33, 53, 104

P

PAM 211
PAR 33, 52, 111
Parallel 23, 35, 52
Parallelism 8, 28, 48, 65
Parameter 53, 96, 130, 168
Parametric 33, 85, 145
Part 3, 10, 40, 163
Partial 100, 177, 192
Partition 32, 48, 91, 120
Partitioned 15, 32, 94, 156
Partitioning 5, 15, 43, 63
Performance 5, 8, 37, 65
Phase 14, 31, 67, 117
Physical 30, 43, 65, 113
PIC 216, 219, 220
PING 57, 59, 183, 190
Policy 35, 97, 134, 146
Population 133, 189, 197, 209
Portion 8, 27, 48, 89
Power 3, 11, 37, 53
Predictability 9, 66, 69
Priority 20, 43, 107
Problem 5, 13, 24, 72
Procedural 4, 54, 132, 163
Procedure 22, 33, 42, 52
Procedure interaction graph 135,
    159, 166
Process 8, 13, 23, 35
Processing 8, 11, 44, 54
Processor 3, 7, 11, 28
Profiling 37, 42, 57, 86

Program 8, 34, 90, 141
Programming 15, 28, 48, 51
Protocol 23, 33, 52, 91
Prototypal 201, 222
Prototype 19, 142, 171
PU 185, 205

R

RAM 122, 138, 165, 204
Range 12, 25, 81, 96
Ratio 74, 92, 154
Real time 9, 27, 50, 155
Register 16, 31, 48, 90
Regular 45, 63, 70, 78
Relation 95, 101, 125, 149
Rendezvous 33, 52, 98, 113
Representation 14, 34, 51, 182
Requirement 24, 68, 221
Round Robin 146, 164, 219
Rule 36, 63, 132

S

SBM 201, 217
SC_CSP_CHANNEL 206, 207
SC_MODULE 206, 207, 216
SC_THREAD 60, 206, 216
SCA 41, 48, 52, 155
Scheduler 150, 216
Scheduling 20, 33, 43, 86
Schematic 110, 146, 216
Scope 33, 105, 176, 199
Semantic 33, 35
SEQ 29, 33, 52, 110
Sequence 26, 54, 76, 107
Set 3, 41, 53, 63
SHARC 67, 165
Shared 30, 117, 144, 209
Signal 7, 31, 63, 110
Simulated 15, 32, 89, 120
Simulation 4, 22, 39, 43
Simulator 29, 43, 90, 134
Size 3, 36, 57, 71
Software 3, 13, 18, 24
Solution 7, 24, 44, 125
Space 3, 15, 25, 45
Specific 7, 15, 36, 43

Specification 4, 14, 22, 28
Specification language 19, 28,
    50, 166
SPP 175, 189, 203, 219
Stack 29, 35
Stage 16, 41, 177
Standard 16, 29, 48, 66
Start 110, 125, 167, 200
State 18, 33, 50, 110
Statement 48, 55, 89, 96
Static 19, 37, 70, 103
Statically 4, 20, 39, 63
Status 30, 110, 150, 222
Step 3, 20, 32, 42
Strategy 4, 72, 142, 172
Stretching 144, 146, 147
Strong 74, 76, 79, 86
Structural 5, 63, 74, 172
Structure 31, 48, 56, 71
Subset 28, 52, 82, 104
Suite 5, 64, 81, 159
SUN 122, 128, 158, 163
SW 4, 13, 85, 200
Switch 102, 146, 158, 196
Synchronization 37, 42, 59, 184
Syntax 30, 56, 102, 108
Syntax tree 34, 56, 93
Synthesis 15, 22, 49, 71
System 3, 12, 17, 23
System design exploration 5, 43,
    61, 135
System level 5, 19, 28, 39
SystemC 47, 53, 172, 199

T

Target 13, 19, 35, 44
Target architecture 13, 30, 44, 135
Task 5, 20, 87, 126
Technique 20, 93
Technology 3, 15, 25, 64
Template 33, 91, 104, 114
Term 15, 58
Test 15, 32, 64, 83
Test bench 32, 208, 216
Time 9, 19, 50, 87
Time stretching 144, 146, 150

Time to completion 179, 201, 220
Timing 9, 32, 42, 53
TO(H)SCA 41, 48, 51, 55
TOHSCA 37, 51, 87, 120
Tool 33, 42, 72, 97
TOSCA 27, 29, 31, 88
Tradeoff 14, 66, 91, 125
Transfer 16, 31, 48, 59
Transformation 30, 43, 180
Translation 16, 32, 103, 104
Tree 34, 55, 93, 110
TREF 137, 164, 169, 170
True 44, 87, 102, 176
TTC 201, 213, 221
Type 18, 45, 73, 117
Typical 31, 42, 66, 128

U
UML 51, 210
Unfeasible 193, 194
Unit 30, 66, 110, 175
User 5, 24, 73, 142

V
Validate 15, 81, 97, 159
Validation 15, 82, 124, 218

Value 58, 62, 74, 96
Variable 56, 90, 105, 119
VCG 55, 134, 155, 164
Vector 21, 68, 98
Verify 13, 32, 42, 158
Verifying 43, 140, 154, 159
Version 28, 32, 52, 185
VHDL 16, 33, 50, 108
View 14, 47, 102, 196
VIS 29, 35, 165

W
Waiting 107, 117, 151, 216
WCTTC 220, 221
Weak 75, 77, 79
Weight 11, 137, 143
WHILE 29, 52, 104, 120
Wide 70, 93, 175, 190
Width 65, 70, 91, 110
Work 3, 22, 51, 88
Worst case 9, 90, 104, 220

About the Author

Luigi Pomante has received the *"Laurea" (i.e. BSc+MSc) Degree in Computer Science Engineering* from "Politecnico di Milano" (Italy) in 1998, the 2nd Level University Master Degree in Information Technology from CEFRIEL (a Center of Excellence of "Politecnico di Milano") in 1999, and the *Ph.D. Degree in Computer Science Engineering* from "Politecnico di Milano" in 2002. He has been a *Researcher* at CEFRIEL from 1999 to 2005 and, in the same period, he has been also a *Temporary Professor* at "Politecnico di Milano". From 2006, he is an *Academic Researcher* at *Center of Excellence DEWS* ("Università degli Studi dell'Aquila", Italy). From 2008 he is also *Assistant Professor* at "Università degli Studi dell'Aquila". His activities focus mainly on *Electronic Design Automation (EDA)*, *Electronic System-Level Design (ESL)* and *Networked Embedded Systems* (in particular *Wireless Sensor Networks*). In such a context, he has been author (or co-author) of near 100 articles published on international and national conference proceedings, journals and book chapters. He has been also reviewer and member of several TPCs related to his research topics. From 2010, he has been in charge of scientific and technical issues on behalf of DEWS in several European and national research projects.

(luigi.pomante@univaq.it; luigi@pomante.net)

Lightning Source UK Ltd.
Milton Keynes UK
UKOW06n1212100117

291770UK00005B/50/P

9 788793 379381